AGAVE SPIRITS

"*Agave Spirits* by David Suro Piñera and Gary Paul Nabhan is a manifesto. It takes a stand against Blue Weber agave, or agave tequilana, and the spirits made from it. The authors . . . [offer] a lively and compelling journey through a large swath of Mexico, highlighting the history of agave and mezcal in Mesoamerica going back 10,000 years; the value of the plant's versatility, taste, nutrition; and its place in Indigenous culture. They offer a positive spin on the future of mezcal, and also make the case for distilling having been used in the region long before the Spaniards showed up."

—Florence Fabricant, *New York Times*

"Mr. Nabhan, an ethnobotanist and naturalist, and Mr. Piñera, a restaurateur, serve as guides to the complex making of mezcal and lead readers through various steps in the mezcal-production process, from growing and harvesting to mashing, fermenting and distilling. But they also aim at a far-reaching overview of the precarious place in which mezcal now finds itself after an extended and harried history." —Brian P. Kelly, *Wall Street Journal*

"Talking about agave with Suro Piñera is akin to discussing the Bible with the pope: His knowledge is so deep and intuitive that it draws you in, even if you consider yourself agnostic, or perhaps (shudder) more of a gin drinker."

—Maddy Sweitzer-Lammé, *Food & Wine*

"The craft of agave spirits is part science and part art, but always about embracing where we come from to understand where we are going. This

book, as with David Suro Piñera and Gary Paul Nabhan's life's work, is a true celebration of how a community's respect for its biodiversity can create liquid magic."
—José Andrés, chef and owner,
José Andrés Group and Oyamel Cocina Mexicana

"*Agave Spirits* invites the reader to a place of wonder and learning all at the same time. It is a captivating journey into the past, present, and future of artisanal mezcals. It's a book about Mexican history and archeology, plant botany and climate change, the promises of mezcal and its kindred spirits, with prose that leaps from the page."
—Pati Jinich, chef, author, PBS host of
La Frontera and *Pati's Mexican Table*

"With agave spirits clearly guiding, Gary Paul Nabhan and David Suro Piñera have blended fascinating science with a loving homage to a mystical plant and the wizards who tend it and extract its juices—and a timely plan for safeguarding its native origins, so we all might keep imbibing."
—Alan Weisman, author of *The World Without Us*

"A thoroughly fascinating dive into all that goes into the spirits, revealing an elaborate web of biological and cultural entanglements, with rich biodiversity in terms of plants, microbes, and even animals, all mediated by skilled human practitioners, carrying out (and refining) cultural traditions. The authors also show how this biodiversity as well as cultural diversity is threatened, concluding with a call to action."
—Sandor Ellix Katz, fermentation revivalist and author of
Fermentation Journeys, *The Art of Fermentation*, and *Wild Fermentation*

"Gary Paul Nabhan and David Suro Piñera combine years of expertise, research, and devotion to capture the artistic intricacies of an ancient drink now threatened by greed. *Salud* to a remarkable achievement and to the rescue of Mexico's emblematic drink."
—Alfredo Corchado, author of
Midnight in Mexico and reporter for the *Dallas Morning News*

AGAVE SPIRITS

The Past, Present, and Future of Mezcals

GARY PAUL NABHAN

and

DAVID SURO PIÑERA

ILLUSTRATIONS BY RENÉ TAPIA

W. W. NORTON & COMPANY

Independent Publishers Since 1923

For information about permission to reproduce selections from this book, write to
Permissions, W. W. Norton & Company, Inc., 500 Fifth Avenue, New York, NY 10110

For information about special discounts for bulk purchases, please contact
W. W. Norton Special Sales at specialsales@wwnorton.com or 800-233-4830

Manufacturing by Lakeside Book Company
Book design by Chris Welch
Production manager: Anna Oler

ISBN 978-1-324-07610-0 pbk.

W. W. Norton & Company, Inc.
500 Fifth Avenue, New York, N.Y. 10110
www.wwnorton.com

W. W. Norton & Company Ltd.
15 Carlisle Street, London W1D 3BS

1 2 3 4 5 6 7 8 9 0

Dedicated to all hearts and souls in the agave spirits
industry and to the memory of three pioneers of agaves and
their place in our cultures,
Howard Scott Gentry,
Efraím Hernández Xolocotzi,
and Tomás Estes

Contents

PART TWO:
THE FUTURE OF THE HUMAN-AGAVE SYMBIOSIS

AGAVE
SPIRITS

David Suro and Gary Nabhan in pulque agave in Guanajuato

Prologue

We are sitting across a wooden table from one another, sipping mezcal. It is late afternoon, the air is fresh, and the sun casts a golden light on the patio of a small palm-thatched restaurant near the southern Jalisco-Colima border. Plantations of a dozen kinds of "century plants"—each with its own hue, heft, and silhouette—mottle the terraces on the ridges above us.

As is often the case when we come together, our discussion turns to the spirits we are savoring—of their texture and taste, and their entrenched yet fragile place in the world of spirits. We invite you to join us in this discussion and to savor at our side the stories and values that make these sorts of agave spirits so exceptional . . . and so vulnerable.

Agave spirits are among the most complex of any distillates. An entire symphony of sensations and scents are embedded in each bottle, as mezcals, ancestral tequilas, and other agave distillates benefit from remarkably rich interactions among succulent plants, fermentation microbes, pollinators, and ancient cultural traditions. Let us immediately alert you to some recent discoveries that underpin our argument: Each of the dozens of wild yeast and bacterial strains found in traditional fermentation vats facilitate the expression of more aroma and flavor volatiles from properly matured agaves than those that can ever be expressed through a single beer yeast working on a single annual grain crop or grape variety. As microbiologist Anne Gschaedler-Mathis has reminded us, "It's not just beer yeasts and common anaerobic bacteria in the fermentation vat; we find protozoans and nematodes as well!"

We will slowly walk you through the fascinating microbiology, invertebrate biology, biochemistry, and plant physiology of agave spirits later on, but for the moment, trust us: Mezcals make other distilled spirits seem like Iowa's monocultures of genetically engineered corn!

If we are correct that such interactions among aged agaves, microbial cultures, and traditions imbue these distillates with a broader flight of fragrances and flavors than we may experience when sampling any other liquor, why isn't that fact already widely appreciated?

For us, it seems there is an entire world to explore within every bottle of agave spirits. But if that is as true as we attest, why then are mezcals and their rustic kin too often unsung and undervalued in most of modern America? Not by every American, of course, but who can deny that many individuals express trepidation when offered their first shot of tequila? Or who among us has not witnessed someone tipping up a cheap bottle of *mixto* tequila and throwing back a shot like a capful of antiseptic mouthwash? For a snapshot of tequila's reputation in popular culture, look no further than the nearest baby boomer's face when someone says "Tequila!" (and then get out of the path of the line dance).

So why is it that so many Americans and Europeans—and yes, Mexicans, too—still perceive tequila as the only—and therefore the best—of the agave spirits, as if it is without peer, when most *mixto* tequilas produced are peered in quality by bathtub hooch, and some small-batch mezcals are truly nonpareil? Join us as we navigate that and a dozen other paradoxes of the spirits world.

We have come together to delight in the diversity of agave spirits. But we are also determined to "deconstruct" and demolish that still-widespread assumption that tequila has some exceptional status among agave spirits. Tequilas of every category are produced from just one species of more than two hundred distinctive kinds of agave, and that blue-hued clone is by no means the tastiest agave in the pack. If we put to rest that exceptionalist attitude regarding tequilas that we see in ads and on billboards, we might remember that many are simply "fair-to-middlin'" spirits historically called *vino mezcal de Tequila*. For centuries the particular agave spirits produced by *tiqulios* or *tequiltecas*—those living in the shadow of the Tequila Volcano in Jalisco—were regarded as just one of the many *vino mezcals* of comparable quality: *zotol, zihuaquio, yocogihua, tuxca, turicato, tauta, tlahuelompa, sikua, raicilla, quitupan, lechuguilla, jarcia, huitzila, chichihualco,* and *bacanora.*

Today, we regard mezcal as the "big brother or sister" in the family of agave spirits, while tequila has become the prodigal son. Spirits like *comiteco* of Chiapas and *bingarrote* of Guanajuato might well be considered the lost cousins of *vino mezcal,* which few North Americans have yet to recognize, let alone savor.

You might say that the two of us first came together out of our shared love for the symbiosis between bats and agaves, and the mutualistic interactions between mezcals and humankind. But we have also endeavored to track down all the lost siblings, cousins, and more distant kin in the agave family that are worthy of your (and our) attention. Our hope is that these artisanal sprits might someday delight you in ways that a Waring blender margarita made with a cheap gold tequila cannot.

So if you are willing to take us on as your trusty trail guides through the Land of the Agave Spirits, perhaps you should know a bit about each of us, what we collectively value, and how each of us came to honor the delicious diversity of agave distillates.

We are just two of the many mezcal historians, scholar-activists, and entrepreneurs who have been promoting and tracking changes in the agave spirits industry for decades. David grew up in the heart of Tequila Country, living his early life in Guadalajara, surrounded by thousands of hectares of the blue tequila monoculture. As an adult based in Philadelphia, he was first engaged as a Mexican restaurateur, and then as a spirits distributor in the United States. Through Suro Imports and Siembra Spirits, he collaborated to craft both ancestral tequilas and artisanal mezcals.

One of David's unique roles is as founder of the Tequila Interchange Project, which brings together chefs, bartenders, distributors, researchers, and journalists on field trips to see firsthand the interactions between nectar-feeding bats and the wild agaves they pollinate in order to protect and restore these vital ecological interactions that anchor the mezcal industry. He has long worked with mezcal makers and scientists to promote better policies in both the US and Mexico to benefit the many stakeholders in this unique Indigenous industry.

Gary is the great-grandson of a Lebanese refugee to Mexico, as well as the grandson and nephew of bootleggers of anisettes like araq. While Gary was still in his twenties, working under the famed plant explorer and agave taxonomist Howard Scott Gentry, he began his pilgrimages to Mexico to meet and interview bootleggers in the Sierra Madre Occidental. It was there that he first witnessed the overharvesting of agaves and the impact on bats while assisting Donna Howell with a *Nova* science documentary on the topic. Along with bat ecologists Howell, Ted Fleming, and Rodrigo Medellín, he was among the first to alert biodiversity advocates to the need to conserve such beneficial mutualisms. As an agroecologist and ethnobotanist, he coauthored the first book on tequila that warned of the consequences of its narrow genetic base. As a plant conservationist and contemplative garden designer, he personally cultivates more than forty species of agave in his "torture orchard" to test their adaptations to climate change near the US/Mexico border.

Independently, both of us developed high regard for the rich fragrances, flavors, textures, and fruitful cultural traditions associated with agave spirits. We also treasure our time with the many families—Indigenous or otherwise—who retain traditional ancestral knowledge about how to craft great mezcals.

Both of us respect and defend the many Mesoamerican cultural traditions that surround and enliven the drinking and "thinking" of mezcal. Mezcal has not evolved in isolation from all the many other foods, beverages, and rituals linked to agaves. We wish to see them neither culturally appropriated nor diluted. Mezcal is not a fad. It is a patrimony.

In fact, mezcal is an iconic element of the ancient Mesoamerican cultural heritage that was recently recognized by UNESCO as a gastronomic tradition of global significance. The culinary preparations and wild fermentation practices of Mexico—invoking agaves, cacti, damiana, maize, tropical fruits, and sotol—have resulted in an astonishing number of unique probiotic ferments and distillates that are celebrated worldwide.

That said, we also value the tenacious dedication to their craft that most *mezcaleros* embody—especially those of Indigenous descent—even though they still suffer from levels of poverty twice that of Mexico's national average.

We recognize that most of the *mezcaleros* in Oaxaca's sixteen Indigenous groups are subsistence farmers or day laborers who have limited means to acquire sufficient land and capital; they are seldom offered a place at the table when political decisions are made that impact their livelihoods. When logistical difficulties or language barriers leave them out of the decision-making processes managed by the Mezcal Regulatory Council, it further keeps them from becoming the primary beneficiaries of the income generated by the sales and fame of their exceptional products.

Whenever we join together with *mezcaleros*, bartenders, and consumers of agave spirits, we reaffirm the values embedded in the simple principles of "good, clean, and fair" that aim to benefit those who bring us our daily drink, not just our daily bread. Inspired by the Slow Food movement, we aspire to dwell in a world where we all of us have the option to affordably drink agave spirits without regret, oppression, guilt, or bad hangovers!

Both of us love the beautiful diversity of the mezcal-producing species variously known as maguey, agaves, or century plants. We wish to see their rainbows of colors, splendor of shapes, and entourage of aromas not only survive but thrive. Agave spirits are not only beautiful to behold, terrific to taste, and arresting in their lingering fragrances; they carry with them a rich and meandering history.

Both of us are concerned for the health of the land and the members of the working class who steward agaves in the field, in the distillery, and in the cantina or bar. We hope to see all of them work for livable wages, in pesticide-free environments, with dignity and well-being.

Both of us care for the *maestras mezcaleras* and *maestros mezcaleros* in the distilleries we have come to know and love, admiring their knowledge, skills, artistic talent, and ethics.

And both of us love the art of good storytelling that we have heard expressed by bartenders, distillery ambassadors, and historians of mezcal culture. Good mezcal is not measured merely by the proof of its alcohol or its price, but by how memorable its tastes and its stories can be.

For us, the pleasures, promises, and problems connected to mezcal and its kindred spirits are all part of the larger story of why we need to heal our planet in time to prevent many species and cultures from certain doom. With accelerating climate change, water scarcity, and habitat destruction, many agaves facing the aggravated risks of extinction—along with other imperiled plants and animals in the same habitats—could be lost forever within the next few decades. Over half of all the 215 species of agaves described to date are among those million species now threatened on the dry, wrinkled face of this earth, the exquisite home that some of us now call Planet Desert. Many of the other thirty-five "mystery" agaves whose taxonomic status is yet to be determined are no doubt rare, threatened, or endangered as well.

That's the bad news, but the good news is that desert-adapted plants like agaves, sotol, prickly pear cacti, pineapple, *mezquite*, and desert oreganos are the kind of steady, slow-growing plants that we need to bring into a new "slow agriculture" to slow down climate change.

By coining the term *slow agriculture*, we envision a resilient, regenerative means of producing the healthy foods, flavorful beverages, effective medicines, strong fibers, and cost-efficient fuels that future generations will require. We are now convinced that agaves and other succulent plants will help our descendants weather a heat-stricken, evermore thirsty world. This work carries with it a certain irony, for our contemporary society is belatedly recognizing what pioneering agave scientists like Howard Scott Gentry and

Efraím Hernández Xolocotzi, and their cohort of colleagues, first realized during the Dust Bowl and Great Depression. To distill the wisdom of the early *maestros mezcaleros*, let us affirm that:

Agaves and other desert plants like prickly pears and sotols are not just pretty faces, voluptuous shapes, or enticing sources of inebriating drinks. They are essential workers in nature's services. They produce the most nuanced spirits in the world. Agaves and other succulents can help us adapt to and survive the daunting challenges of global climate change and catastrophic weather events.

Agaves and their kin are not only bizarrely diverse in their architectural forms, flavors, fragrances, and uses, but they sustain a vast diversity of bats, carpenter bees, fruit flies, hummingbirds, larvae, root microbes, shrews, and silverfish in their flowers and in their flesh.

These succulents can produce more edible or drinkable biomass on less water than most conventional crops like potatoes, wheat, rice, and maize can muster.

As succulent perennials, agaves can sequester more carbon to offset fossil fuel emissions than can most annually planted edible or industrial crops.

Agaves produce far more than just tequila and mezcal: food and fiber products, medicinal remedies, structural and architectural materials, "nectar" or molasses, and parchment-like waxy leaf cuticles called *mixiotes* used as a wrap in pit barbecues. They support all manner of ceremonial and ritual uses, ones too numerous to circumscribe.

Roasted agave *pencas* (leaves) and *mezontes* (hearts) render "slow-release" foods that slow down the digestion and absorption of sugars in a manner that reduces blood sugar levels and increases insulin sensitivity of the growing number of people who suffer from adult-onset diabetes.

Agaves also contain substances that fight inflammatory diseases, thereby reducing the risk of the emergence of certain cancers and kinds of atherosclerosis.

Agaves are exquisitely beautifully examples of mandala-like radial symmetry and mathematically elegant spiral designs, such that they are as good to meditate upon as they are to drink or eat.

These long-lived "century plants" teach us patience, resilience, perspicacity, and tenacity. Oaxacan Vicente Reyes speaks of the Way of Maguey as a spiritual path: "The Maguey Path is continuous and infinite and like the relationships between life and death, there can be no duality; we can also see this infinite cycle that is also destiny among the stars and El Maguey who drinks the morning dew through her stalks."

It is time that humankind follow "the Way of Maguey" at this critical moment in our history and the history of our Planet Desert. In the following pages, you will learn to talk agave, think like an agave, endure like an agave, and change the way we as humans interact with agaves so that we can carry on and collaborate on this ever-drier planet.

Let us transform and restore it back to the Planet of Sensuous Symbiosis. Can't we all drink to that?

PART ONE

MEZCAL'S
HISTORIC LEGACY

Chapter One

※

THE FAMILY OF AGAVE SPIRITS

For many of us, our initiation into the culture of tequila, mezcal, and other agave spirits was love at first sip. We intuited that we were entering into an endlessly fascinating domain of Mesoamerican culture, agriculture, folklore, and pleasure, though we could not initially taste or see all the profound dimensions of the agave spirit world.

During those first tastings, we could not have possibly known that bottles of mezcal contain the benefits of more microbial and plant diversity than any other kind of spirits.

How could we immediately fathom that growing, roasting, crushing, fermenting, and distilling the hearts of agaves could generate such a bright constellation of tastes, textures, and aromas that burst like fireworks into our cups and glasses?

How could we understand that six million years of agave evolution and adaptation to harsh conditions on Planet Desert is condensed into every clay bowl, *jicarita* gourd cup, or shot glass we drink from?

How could we sense the amount of time and ingenuity it took to develop the slow agricultural and gastronomic traditions that Indigenous cultures of Mesoamerica and Arid America use to cultivate, ferment, and distill agaves?

By referring to *slow agriculture* now and then, we wish to promote the investment of the patient capital of natural processes and stewardship by humans to generate food and fiber over the long haul. Such an agriculture builds up the soil's moisture-holding capacity and carbon reserves to assure future harvests as well. A century plant in an Indigenous farmer's multi-cropped field has exemplified that kind of investment in the earth over many millennia.

By *slow gastronomy*, we mean the investment in slow fermentations of decades-old century plants into delicious, nutritious foods and probiotic beverages that are then distilled and cured over many months, if not years.

After celebrated spirits journalist Emma Janzen listened to the craft distillers whom she had met in *mezcalerías* throughout Mexico, she realized, "Many a *maestro* has exclaimed, *mezcal tastes like time,* because the plant takes so long to mature, and the production process is slow and laborious. Patience is key to making good mezcal."

Yes, deep time. Agave evolution itself has been a slow, patient process. The first agaves and their closest kin (as we recognize them today) diverged from asparagus-like lilies six to seven million years ago. With the emergence of the towering cordilleras of the Sierra Madre Occidental and Trans-Volcanic Belt in present-day Mexico, many agaves specialized their morphologies for a life in drought-prone areas called "rain shadow deserts," such as the Tehuacán Valley. That is where they took on their iconic forms with sword-shaped leaves, succulent (water-conserving) core tissues, and gigantic flower stalks.

Looking much the way that agaves appear to us today, these hardy but primitive succulents began to slowly adapt to different kinds of arid landscapes where varying rock types, micro-climates, and migratory pollinators further influenced their evolution.

Then, during the Pliocene and Pleistocene—around 2.7 million years ago—the evolution of agaves changed tempo. The rate of differentiation and diversification among agaves sped up, perhaps as a result of two factors: geographic isolation and interactions with pollinators. The isolation occurred as geographic barriers were created by the last extensive glaciers and their ultimate retreat. Novel species emerged with candelabra-like flower stalks and protruding blossoms suited to visitation by bats. This divergence of the agave genome into altogether distinctive forms is what biologists call *adaptive radiation*. Some agave species invested energy in erecting the tallest flower stalks known in the plant world—48 feet (15 meters) in height! They were rewarded by having more kinds of pollinators successfully fertilize their flowers, resulting in higher seed set and reproductive success.

Over time, these evolutionary innovations led to a rainbow of leaf forms and colors, and shapes of flower stalks that could hold hundreds or even thousands of individual blossoms with musky fragrances. Individual rosettes of leaves might grow for thirty-six to seventy years before partaking in what evolutionary ecologist Luis Eguiarte calls "suicidal reproduction," but the clonal offshoots of one seed-borne mother plant might live on for centuries or even millennia!

That's right: The very best mezcal agaves have been brought to you by "patient capital"—long-term investments—as well as austere aridity and nocturnal, nectar-licking, pollen-nuzzling bats and moths.

Today, the evolutionary trajectory of more than 215 agave species is facing another singular moment of change, one characterized by crisis, conflict, and potentially cataclysmic collapse. After millions of years of diversification, a few species are becoming monotonously uniform, and are dominating the total number of agaves found out in the landscape. We must now decide whether mezcal and other rare agave distillates should continue to follow most brands of tequila down the perilous path of industrialization, or stay true to their diverse history and deep cultural traditions.

You have probably noticed that we have been using the word *mezcal* a lot, but what concerns us here is an entire fleet of 100 percent agave spirits that have evolved in Mexico for centuries, not only those regulated under the official legal definition in the Denomination of Origin for mezcal that guides the industry. In fact, there is a vast clan of agave spirits in a dozen states that have not been officially accepted as mezcals by their current legal definition and regulatory protocols, but that is what most Mexicans call them nonetheless. Up until the 1970s, the scientific term *agave* was simply not part of the spoken word for most *mezcaleros*, nor for the majority of drinkers in Mexico and beyond. As *maestro mezcalero* Miguel Partida told us once on the plaza of Zapotitlán de Vadillo, Jalisco, "In the old days, my grandfathers never used the word *agave*. The plants were called *mezcal*, not the drink. The drink was simply called *vino*. That is what's so crazy about how the Regulatory Councils want to legally define the term *mezcal*, to prohibit its use in other ways. It ignores the way campesinos have used the word for centuries, and we can no longer use that term on our labels like we used to!"

So, forgive us if we now and then use the term *mezcal* in a broader manner, as an equivalent of the many kinds of agaves growing in Mexico, or of the many 100 percent agave distillates historically made from these century plants. In fact, tequila comes from just one of dozens of regional mezcal-making traditions, as suggested by its old name, *vino mezcal de Tequila Jalisco*. Somehow along its wayward journey, tequila lost most of its connections to its ancestral roots, going astray.

That one fact alone—tequila's singular isolation—is key to understanding the current state of the agave spirit world. Before we steep your being in a bathtub of artisanal mezcals, we'd like to caution you that many agave spirits are now heading down the same superhighway that tequila has been barreling down over the last four decades.

Traffic on that superhighway is now dominated by just five brands—Jose Cuervo, Patrón, Sauza, 1800, and Hornitos—that all sell more than a million cases each year, nearly three-quarters of them destined for the US marketplace. Aside from Cuervo, which is owned by the Beckman family that splits their time between Mexico and the United States, the other brands are all owned by multinational corporations such as Bacardi USA, Beam Suntory, and Proximo Spirits. Other large corporations like Diageo, Luxco, Brown-Forman, Gallo, Sazerac, and Campari are hot on the heels in the race to dominate the global beverage market.

By the way, have you noticed many Mexican-based corporations among those frontrunners? Nope. Because they are not there. Might that suggest to you that adherence to tradition and patrimony is not on the top of every executive's list? Mexican-owned tequila brands, like Fortaleza, Tapatio, Cascahuín, 7 Leguas, and G4, are no less enjoyable, they just lack as much fast capital to accelerate and expand their marketing and distribution.

Whenever tradition is at stake—as it clearly is today—some will argue that change is inevitable, inexorable, and inescapable. They will claim that you can't slow or stop "progress," and that any old-fashioned traditions associated with food, drink, dance, music, or other ancient art form are not even worth fighting for. They maintain that human enterprise must endeavor to become ever more competitive, cost-efficient, and profitable, riffing and ripping off once-obscure products, making them commodities, available at a click to billions of people.

If consumers only want to take part in some new trend, why hold on to tradition for tradition's sake when it seems that it just gets in the way? In the case of many unsung agave distillates—like *bacanora, bingarrote, chichihualco, comiteco, excomunión, huitzila, jarcia, raicilla, sikua, tasequi, tauta, tlahuelompa, turicato, tuxca, zihuaquio,* and *zotol*—there are many traditions we risk losing because someone deems them obsolete.

Where agaves grow best in Mexico and the Desert Southwest, we are also losing the culinary preparations of eating pit-roasted agave hearts, of drinking the probiotic beverages made from fermenting them, of patiently peeling off the leaves of the waxy *mixiote* cuticles used in pit-roasting or barbecuing, and of toasting larvae and insects harvested from agaves to grind into flavored salts.

With these gastronomic practices goes the loss of orally transmitted sayings, harvesting skills, and recipes formerly passed on for hundreds or thousands of years by Indigenous agave cultivators, cooks, and *curanderas*. They knew agaves and other plants from the inside out, with a depth of familiarity that few scientists muster today.

As Vicente Reyes, the Oaxacan-born poet has noted, "The knowledge that

El Maguey holds is ancient; it has been and will be part of Indigenous people's wisdom. It is not the stuff of books or information gathered in data that
can be understood by the rational mind."

And so, there is much at stake when it comes to which road the makers
of agave spirits choose to take over the next two decades—quality spirits,
satisfying livelihoods, deeply rooted cultures, and the diversity of life in the
Americas. As food journalist Mark Bittman recently put it, "How we eat and
drink is how we treat our planet, our neighbors, and ourselves."

Should we lose lesser-known Mexican spirits, we will sacrifice a healthy,
diverse future for the agave spirit world. We must maintain diverse and
dynamic cultural traditions that foster the emergence, continuance, or
resurgence of exceptional and irreplaceable beverages. These beverages are
deeply steeped in ancient traditions. Maintaining these traditions depends
on seasoned practitioners with immense traditional knowledge and skill,
who know how to plant, when to plant, when to trim, and how to prepare
agaves for food, ferment, fiber-woven bags, and rugs, ropes, rites, medicine,
and more. They also know to safeguard the very ecological and cultural conditions required to sustain what we call the *mezcal-humankind mutualism*
into the future.

And yet, you would never know that your choice of what agave spirit to order
in a bar or pull off a liquor store shelf had any relationship to these bigger
issues if you simply took ads for tequila at face value. Since the mid-1970s,
every tequila on every shelf in every bar in the world has been made from
homogeneous clones of just a single agave variety out of the hundreds that
exist: the *tequilana azul* cultivar of *Agave tequilana*.

Today, tequila is as impoverished in its plant diversity, microbial diversity, and heterogeneity of aroma, flavor, and fragrance as any crop in the
world that is used to distill alcohol. A single liter bottle of 100 percent agave
tequila now takes 11 to 15 pounds (5 to 7 kilos) of genetically homogeneous
agave, consuming 35 pounds (16 kilos) of firewood and 65 gallons (nearly
250 liters) of water on its journey through production and processing. As *Eat*

Less Water author Florencia Ramirez estimated in 2017, "Annually Americans sip, gulp, or mix 151 million liters of tequila, with a water footprint of nearly 10 billion gallons, enough to fill 17,845 Olympic-size pools."

It is as if someone waved a magic wand, and *presto!*—all the diversity and frugality embedded in tequila's deep history were drained out of every bottle or spilled onto the earth.

※

We all recognize that there is more to a bouquet than one type of flower. So too, as anyone who has sipped a quality mezcal can attest, there's a lot more valuable variability in agave distillates than in those cheap *mixto* tequilas made from blue agave, streamlined strains of laboratory-bred yeast, and mounds of added sugar. There is also much more to agave agriculture than the blue desert of tequila monoculture now found on hundreds of thousands of hectares in five Mexican states.

In Jalisco alone, other incredible varieties and distinctive species of agaves like *bermejo, listado, moraleño, pata de mula, sahuayo, sigüín,* and *zopilote* were once common. That's because they were all commonly used in the historic elaboration of *vino mezcal de Tequila,* the precursor of modern industrial tequila. That's right: The blue tequila clone was not the only kid on the block a century ago in central and northern Jalisco.

To her credit, agave agronomist and visionary Ana Valenzuela has not only rediscovered these varieties on the margins of tequila production, but was the first to attempt to conserve them in "nursery gardens of endangered diversity." Ana has proposed a fresh new agenda for maximizing their value. At the same time, her historic detective work has made it abundantly clear that propaganda about the primacy of *tequilana azul* for making the best of the agave spirits is elitist hogwash. It ignores the fact that until the 1960s, the *tequilana azul* cultivar was just one of eight different agaves traditionally used in *vino mezcal de Tequila* for centuries.

Of course, just as you may find some white roses to be beautiful, you may find several noteworthy tequilas made from this single agave cultivar. Nevertheless, the reasoning behind the exclusion of other agaves was not based on a desire for better quality control or exceptional flavor; it emerged out of a materialistic need (or greed) for supply chain efficiency, uniformity, exclusivity, and expediency.

✺

To be sure, the *tequilana azul* cultivar did indeed have a few immensely valuable qualities suited to economic growth. It grew rapidly in most of Jalisco and adjacent states, for its most precocious clones mature in less than half the time required for other agaves. It can be prolific in generating *hijuelo* sideshoots of the mother plant (also known as "seeds" by some planters) to be used in propagation. This is classic vegetative propagation, like you might do in your garden for chives, garlic, or onions, using genetically uniform propagules trimmed from the base of the mother plant rather than true "seed" harboring hidden genetic variation. In addition to these agronomic characteristics, the carbohydrate content of its sweet fructans is exceptionally high, while its levels of indigestible fiber and bitter, sudsy saponins are remarkably low.

In essence, you can squeeze a lot of sugar out of a blue tequila plant over a relatively short period, for it may be harvestable in as little as five or six years from outplanting. Tequila's life cycle is shortened so much by biotechnology that one wonders whether it should still be called a "century plant" at all. If you feel you can tolerate what some activists call "plant infanticide," then industrially processed tequila is for you.

While the pros of increased efficiency are compelling, was it really worth stopping the use of seven other agaves in the production of tequila in the name of purity? Would California florists suddenly stop using anything but Iceberg floribundas in bouquets just because they grow well in California? No. So why should Mexico, a country that prides itself on being a global hot spot of both wild and cultivated plant diversity? Nevertheless, Mexican's big businessmen opted for the Faustian bargain: a Denomination of Origin for tequila, administered from the top down, with the singular goal of sucking money upward.

✺

A Denomination of Origin (DO) is a government-refereed system of protections for certain specialty and artisanal products. These place-based protections are one way to circumscribe distinctive food or beverage traditions of a particular region or a nation. Other legal tools for protecting place-based beverages are Appellations, Geographic Indications, and Traditional Specialties.

Whether wine from the Burgundy region of east-central France or maple syrup from sugar bushes growing on the schists of northern Vermont, food products have what's known as terroir—the unique taste of a particular place—and the scents of a seasoned culture are bursting from them.

The goal, in theory and usually in practice, is protection of the quality and distinctiveness of the products, solidifying the reputations of the products, and through that, safeguarding the livelihoods of those who produce a food or beverage according to the traditions of a particular place or culture. Hypothetically, if Taylor Ham could rebrand and widely distribute their products as Iberian Ham, the Iberian Ham category would lose its reputation for excellence, no one could be sure that a product labeled as Iberian Ham would be of the highest quality, and its true producers in Iberia would be hung out to dry.

What precipitated the tequila industry's sellout to quantity over quality? In the 1970s, as the demand for margarita slushies soared on a steep curve, Mexican capitalists decided it was time that the DO for tequila be created. The Mexican government published the resolution for the DO in December 1974, and four years later the Appellation of Origin for tequila was formally registered before the World Intellectual Property Organization (WIPO), officially initiating its international protection. For the labeling of a product as "Tequila" under this place-based DO, its production was restricted to 160 municipalities found in just five states that became the declared territory of Tequilandia in parts of northern and central Mexico.

Ironically, these various states and *municipios*, however, fail to share an identifiable terroir for their spirits. Neither do they share a distinctive cultural tradition of artisan processing, as *olla de barro* mezcal producers do in Oaxaca. By objective criteria, the tequila producers in the five states lacked the very geographic and cultural cohesion demanded of other foods or drinks under most Denominations of Origin.

In short, tequila territory was a vast hodgepodge of disparate natural and cultural features, and simply reflected where the *jefes* running the major Mexican *empresas* involved in the spirits industry owned their properties. For over a century, the tequila elite comprised wealthy Jaliscan families with large haciendas and substantial political pull. The daughters of several of these kingpin players married foreign engineers of equal wealth, who knew how to modernize and upscale their distillation technologies and to move their products by seafaring vessel and rail. Their desire for including a rather small but accessible area of the blue agaves grown in Tamaulipas was spectacularly nonsensical, since there was not a single tequila distillery there for decades; all of it was shipped over to the "big dogs" in Jalisco for processing.

Three decades after the DO for tequila was ratified, a particularly "blue" production area of 135 square miles (35,000 hectares) within Jalisco's valleys got itself declared a UNESCO World Heritage Area. Its banners touted the clunky moniker, "Agave Landscape and Ancient Industrial Facilities of Tequila." The evaluation team members working for UNESCO must have scratched their heads, wondering what actual "heritage" they were charged with "protecting." Why did it include the hyper-modern industrial facilities housing the bottlers, the inulin extraction plants, and the warehouses where pesticides and herbicides were hidden in bottles? Are those part of Mexico's ancient patrimony?

No, they are not. The World Heritage Area for tequila, or Paisaje Agavero, is simply an incongruous mishmash of stodgy factory-like processing plants, warehouses, industrialized distilleries, and agrichemical plants, with tourist-friendly facades—tasting rooms, historical museums, and industry showcases—scattered in between. You hear more diesel trucks than mariachis and see more stores hawking herbicides than local-owned hardware shops offering the wrought-iron *coas,* hoes, and machetes that have long been the tools of the trade.

On the clearest days, you can still see the iconic volcano of Tequila towering over the killing fields, but a haze of diesel smoke, pesticides, and dust from plowed-over fields rises up its flanks. To make matters worse, the intrusion of berry production upon agave fields has placed the entire landscape at risk of being decommissioned by UNESCO.

It was a made-for-television heritage area, as deeply historical and authentic as the Corn Palace in Mitchell, South Dakota, the London Bridge on Lake Havasu in Arizona, or Dollywood in Pigeon Forge, Tennessee, which are now among the most widely visited "shrines" in North America. And it is now being contested as a valid heritage area by UNESCO itself.

If you wish to evaluate the success of the Denomination of Origin for preserving the integrity of tequila, look no further than the dubious reputation of the most widely distributed *mixto* tequilas. To put it bluntly, these sugarcane-saturated facsimile tequilas have not survived Taco-Bellification. Compared to other agave distillates, tequila has become what Velveeta "cheese food products" are to the larger world of artisanal cheeses. For decades, we have fought to defend and protect tequila's integrity. More recently, we have fought to defend the many mezcals under the broad banner of the Consejo Regulador del Mezcal, one of the regulatory councils required by the Mexican government to regulate, guide, and ensure the quality of products in the industry.

To understand why so many Mexican, North American, and European drinkers, distillers, distributors, and scholars are now defending the myriad mezcals against tequila monoculture, we need to take you into Mexico's prehistoric ruins, terraced plantations, traditional fermentation vats, and styles of stills. We will return to the dilemma of the dangerous decline in tequila diversity later on, in Chapter 9, but for the moment, read everything that follows with this in mind: There is much to be appreciative of and awed by in the agave spirit world, but there is also much at stake.

Chapter Two

🌾

DRINKING AND THINKING
LIKE AN AGAVE

A s we raise up our glasses of a traditional distillate of agave, we seldom think that when we drink it, we are taking in the spirit of another being. We do not consciously sense that the molecules, cells, and isotopic signatures of another sentient, sensuous, and spiritual life are becoming embedded in our own.

Yet from the very first scientific naming of these "century plants" or "American aloes" in 1797, they were conferred a special spiritual status relative to other representatives of the plant world. The scientific moniker *Agave* is derived from a cluster of Greek lexemes connoting mythic dimensions, including *agauos*, "noble, illustrious, high-born or saintly," and *agasthai*, "to wonder at, rejoice, exult." We might conclude that early botanists sensed that agaves were not merely beings from the world of plants but were also elevated to the spirit world.

🌾

To know mezcals (as beverages) in the broadest sense of this word, we must come to know the very spirit of agave, for it is the most complex, patient, and resilient of any spirit of the floral kind. As Emma Janzen observed in her 2017 book, *Mezcal: The History, Craft & Cocktails of the World's Ultimate*

Artisanal Spirit, "Mezcal can be distilled from no plant other than those that belong to the botanical genus *Agave.* So, when we imbibe mezcals, we become—metaphorically and biochemically—members of the *Agave* family. In fact, agaves are the only succulents in the world commonly employed to distill alcoholic beverages. So to fathom exactly why so many mezcals embody so much amazing flavor, fragrances, and substance, we must get a sense of the amazing ways agaves have found to grow and survive in a world of stress and uncertainty."

Or as agave distiller and scientist Iván Saldaña Oyarzabál put it in his essay, *The Anatomy of Mezcal,* "Mezcal should make us humble. There are dimensions of mezcal that cannot be grasped by using rational information, history, or biology. Perhaps it is the overall sensory experience that it creates, the state of mind and emotions it can bring on. The shamans of the Zapotec and the Mixé regions use it to travel and think."

The possibility that agaves are both sentient and wise has never been expressed more clearly than in the way *maestro mezcalero* Tomás Virgen said it to us in the town plaza at Zapotitlán de Vadillo, Jalisco, where we sat together below the flanks of the Volcán de Colima: "Agaves are very intelligent. During droughts, they know when and how to slow down their activities and roll up their leaves to conserve water. But as soon as it begins to rain once more, they know how to quickly reopen their shops and catch all the rain in their gutter-like leaves to trigger a fresh spurt of growth! There is a lot that we can learn from them."

When two mezcaleros of international renown both humbly defer to the sageness of a straggly, sometimes spindly, or sprawling plant as a source of intelligence, perhaps the rest of us should take notice! By thinking and drinking like an agave, we might someday learn how to live pleasurably in the water-scarce world of Planet Desert. That Braised New World is just around the corner, and we have barely awakened to its reality.

The linguistic roots of the term *mezcal* may have grown out of the Nahuatl conjunction *mexcalli,* a joining together of *me(tl),* "agave," with *xcal,* "roasted." The roasting of this peculiar plant in firepits is sealed with the suffix -*li*, because that is just what Aztecs did to finish off a proper name if it ends with the letter *l*, as this one does with -*xcal*.

In short, the essence of *mezcal* is the wedding of agave with fire, for you need the energy and aromas from certain woods native to Mexico to roast and distill agaves into something divine. If you don't know the agave itself, and the fire that burns within it, it is hard to know what makes a good mezcal.

Well before the Spanish conquistadores arrived and first witnessed the fields of maguey in Mesoamerica, the use of that Nahuatl word *mexcalli* had spread far beyond the Aztec homelands of central Mexico. A dialect of their language, Nahuatl, was also used as a lingua franca for trade among many other cultures, especially those that exchanged hallucinogenic plants, parrots, cacao, incense, obsidian, turkeys, and turquoise from as far south as Central America all the way up through what we now call the US Southwest.

As this Nahuatl term for agaves traveled far and wide, *metl* was modified to describe many distinctive shapes, sizes, and *sabores* of maguey. Different prefixes were attached to it in order to describe the panoply of agaves that inhabit the lands of Mesoamerica and Arid America. The Aztecs "sentimentally" used the term *centemetl* for the unusually gigantic *Agave atrovirens; hocimetl* for the divinely curvaceous *Agave inaequidens; mecometl* for the silvery swordlike leaves of *Agave americana; metometl* for the fibrous and soapy *Agave lechuguilla;* and *papalometl* for the butterfly-like shapes of *Agave potatorum.*

Of course, there are additional kinds of *metl* as well—as we shall soon see and taste—but each of them reminds us that the Aztec goddess of agaves, Mayahuel, had as many "breasts" as the Hindu goddess Durga had arms. There is not just one dominant species of agave, as there is for maize or the apple. Instead, agaves exemplify a rainbow coalition, a high-spirited congregation, a motley crew of plant organ donors.

※

That said, the term *mezcal* has been overused and abused. It was traditionally reserved for pit-roasted products of agaves. Therefore, it should exclude some kindred spirits such as sotols (from the genus *Dasylirion),* and those tequilas made without pit-roasting agaves, as well as the industrial variety that are infused with nearly as much cane sugar as agave sugars. Neither should *mezcal* from agaves be confused with the poisonous mescal beans of the shrub called *Sophora*, nor with the hallucinogenic mescaline extracted from the peyote cactus called *Lophophora*.

Aside from what particular agave plant is harvested, the kind of wood used in roasting mature agaves in underground pits or brick ovens is fundamental to the fragrance and integrity of *mezcal*. That is because the smoky, caramelized, or faintly charred scents of different mezcals are derived from burning hardwoods that are specially selected for particular aromas: *mezquite*, red oak, white oak, cottonwood, pine, and buckbrush or lilac. Other woods like oyamel firs *(Abies religiosa)* are selected for the vats used in fermentation, or the barrels in which aged mezcals reach their angles of repose. Unfortunately, these trees are becoming locally scarce due to the growing demand for roasting and distilling agave spirits.

To be sure, sweet substance can be extracted from agaves without roasting them over a wood fire—by throwing immature plants into autoclaves and diffusers—but without the same sensory delights or organoleptic effects. As tequila researcher Anne Gschaedler-Mathis confirmed to us in her CIATEJ laboratory in Guadalajara, "Without a doubt, there are multiple consequences of harvesting immature plants in tequila populations." No quick technological fix provides mezcals with the flavor and fragrance that roasting mature plants in a wood-fired oven or pit can offer.

It is easy to feel overwhelmed by all the diversity in the agave spirit world unless you keep one fundamental fact in front of you: *Agave spirits are distinctive relative to other distillates because they originate from a clan of plants like no other.* To comprehend why agave spirits are so distinctive, you must fathom what most agaves are up against in the harsh and uncertain environs where they grow.

Most spirits are derived from grasses (cereals and sugarcane), from grapes and other fruits, or from tubers like potatoes. But agaves are gifted with completely different chemical signatures to the flavors, fragrances, and textures found in their mash or must, signatures unlike anything else in the materia prima employed for elaborating other spirits.

Each time you sip mezcal from a *jicarita* gourd cup or an embossed shot glass, it is dripping with sagas of suffering and survival. You are tasting evidence of the exquisite adaptations agaves evolved to face daunting uncertainties and undeniable stresses as they struggled to survive and reproduce, ones far more severe than those faced by any other plant that we might imbibe as a liquor or elixir. The essence of those struggles will end up in every mouthful of *mezcal, raicilla, bacanora,* or *comiteco* you will drink over your lifetime.

One of the most beloved and widely quoted scientists of the Americas—the internationally renowned ecologist Exequiel Ezcurra—once reminded us that the essence of an agave is a bunch of sweet, syrupy carbohydrates stuck on a flayed, fibrous skeleton that opens its leaves up to the sun. No wonder that over the last ten millennia, humans have chosen to use the ubiquitous agave as a source of fiber for weaving ropes, bags, belts, carpets, and sleeping mats. No wonder that they figured out how to access agaves as an assured source of fermented sugars for drinking. Sweet, sticky liquid energy wrapped around durable structures . . .

That is what agaves are all about, according to the famed plant doctor, Ezcurra: "Over five thousand years, agaves have provided to Mexican populations two key products, fibers from leaves designed to have long parallel nerves, and sugars. Both are a consequence of the necessity to accumulate energy and nutrients at the base of the rosette that will later be utilized for flowering and fruiting."

Of course, there is good reason we call them century plants, even though few if any individual agave rosettes become centenarians. They stay celibate and delay having sex much later than other plants, but then they do it with a bang. Agaves are resilient, patient perennials that can endure hardship over many decades—as if on some extended rite of passage—before they are "released" to perform their rather remarkable reproductive tricks.

When the agave's time comes, it sends up enormous flower stalks, blossoms, blushes while being pollinated, and whenever the consummation of sex is successful, it sets seeds, then invariably dies. *Suicidal reproduction!* (Of course, the analogy is imperfect because the bat and the agave aren't having sex with one another, and the bats themselves do not die. The agave pollen that was moved from the anthers of one flower to the pistils of another germinates on the style of the stigma to grow downward and move into the ovary to reach the ovule, thereby "fertilizing" it.) A big bang, then a whimper out of hearing distance and sight from all.

Yes, agaves literally kill themselves trying to reproduce. They deplete all the reserves of their lifetime just to have one season of sex that is literally to die for. Oh, it is sad but true that each individual agave rosette gets only one fleeting chance at sexual consummation over its entire lifetime.

Well, that is not exactly the entire story. It is tough to discern what an "individual" agave really is. Each rosette on a "mother plant" may have root-like underground branches called rhizomes that extend out through the soil to spin off smaller, younger rosettes. These branches allow the larger network of the same "genet" or genetic individual to live on even after the plant aboveground dies. Each of these offshoots or "ramets" is called an *hijuelo*, and their rosettes will eventually send up flower stalks as well. In short, each rosette of leaves is like the terminal branch of a many-headed plant, the botanical analog of the many-headed water serpent that the Greeks called Hydra . . . or the many-breasted goddess, Mayahuel.

By pruning the offshoots to detach them from the connective tissues of the underground rhizomes, *mezcaleros* can use them for the vegetation propagation of "new" agaves. That is the second most common means by which agaves reproduce and sustain their kind. It has also become problematic, since the overuse of clonal propagation in the field or in the lab is what leads to monoculture.

Their third means is by plantlets called bulbils that emerge from the flower stalk after the flowers themselves have dropped. They are genetically identical clones of the mother plant, just as tiny onions, garlics, or maternity plantlets may be.

In sum, each discrete agave head or *cabeza* can reproduce sexually only once, but the entire genetic individual or clone has been granted the next best thing to eternal life. Its instinct for continual vegetative propagation is nearly as good as reincarnation. The century plant takes the long view of the world around us.

※

As we hinted earlier, it is this capacity for a long life that makes agaves perfectly suited to the investment of patient capital in a kind of *slow agriculture*. Over time, one agave crop can yield far more edible or drinkable biomass per decade than ten different plantings of annual crops can achieve over the same period. As such, forms of agriculture based on cultivating perennial agaves may be the perfect antidote to the quick fix, boom-and-bust cycles of short-lived crops that deplete soils, water reserves, and human capital.

You may already be asking yourself the obvious question: How can agaves sustain their lives for so long under the constant challenges of living in deserts, semi-arid steppes, dry pine forests, barren cliff faces rising above jungle canopies, or drought-ridden subtropical scrublands?

Well, they do so by investing in a set of structural and chemical engineering strategies so efficient and ingenious that they boggle the minds of most architects, biochemists, mechanical engineers, and physiologists. Let us see how they can go far beyond surviving under stressful conditions to comfortably thriving where few other plants can.

※

To think like an agave—to use a metaphor coined by the late Tony Burgess, a prophetic Texan—you must first envision a radiant mandala. That is exactly what a rosette of agave leaves forms, one exquisitely designed to achieve two goals. First, it must efficiently absorb lots of photosynthetically active light without burning itself up. Second, it must harvest scant rainwater to keep its internal tissues moist and succulent. Its radial architecture allows both goals to be achieved.

Like a nautilus shell, a napa cabbage, or a pineapple, what first appears as a mandala of agave leaves is a spiral of fractals dancing around the heart of the plant. At the heart of the agave is a core of meristematic tissue called the caudex. As buds emerge from the meristem, a mathematically precise pattern of swirling is set up that is called the Fibonacci sequence, or "golden spiral."

From this core tissue, the primordium of each leaf emerges unimpeded in its reach toward the heavens, so that over a day's span from sunrise to sunset, it achieves near-optimal absorption of the spectrum of solar energy that fuels plant growth. No leaf ever fully blocks the one below it, because each is slightly offset in a way that allows more sunlight than shade to fall upon it.

And yet, even on the hottest of days, there is enough shingle-like shading of one agave leaf by another as the sun moves past them that they seldom suffer from excessive heat loads. Most of the leaves have a thick, waxy *mixiote* cuticle that simultaneously reflects heat and retains moisture. The various colors and sheens of agave leaves also help the plant reflect excessive solar radiation and heat—the silvery surface of monstrous *Agave americana* leaves is so brilliantly reflective that it repels certain leaf-sucking insects in their attempts to siphon off moisture from the plant.

As scientific illustrator Paul Mirocha reminds us, "Agaves can teach us both math and art, but we in turn are hard-wired in our minds to love their visual harmony, their elegant architectural forms."

When we quietly reflect upon their perfect spiral form for a while, our blood pressure levels may drop, our heart rates may slow to a steadier pulse, and a sense of deep tranquility may emerge from within us.

Now, here is where we get into the colorful drinking habits of agaves. The leaves may look like swords, but they are not designed to be weapons as much as water-harvesting conduits. They are angled and guttered to drain any rain that falls upon the leaf surface, however briefly or intensely, down to the leaf bases and, ultimately, to the roots hidden from the sun. Most of the leaves are hunkered down close to the ground, where they channel rainwater to spill out immediately below them, rather than draining away. That moisture is absorbed by the roots, which are buffered from the evaporative pull on a hot day.

Because some agaves used in mezcal-making have leaves 6 to 8 feet (2 to 2.5 meters) in length, they harvest far more rainwater in a storm than that which runs off the thorny backs of horned lizards or the spiny flanks of saguaro cacti. They are utterly good at being a gutter.

What is more, those trough-like leaves funnel that rainwater runoff down to the roots with remarkable rapidity since most of the receptive roots are within 6 inches below the surface of the soil. In a microburst of summer rains that drops less than a third of an inch of water, this meager supply per square inch of leaf surface is multiplied, as it drains down and accumulates beneath the leaf bases, where root primordia begin to bud out and absorb this moisture within minutes after the onset of a storm.

The sudden stream of water is sufficient for the agave to open shop for a spurt of new growth that can persist another eight days without another drop of rain. The leaves open their stomatal pores and snap into work, making hay (or future mezcal) for as long as the agave's tissues stay wetter than the soil itself.

Desert plant physiologists call this capacity a "pulse/reserve" strategy, for agaves can quickly take advantage of a brief windfall of moisture, and then hold on to it for months. Like Western desperadoes or desert hunter-gatherers, they are supreme opportunists that know how to lay low and beat the heat whenever the risks get too high.

To put it another way, agaves know how to drink responsibly. Plant physiologist Park Nobel determined that succulent agaves consume and use water with eight times more efficiency than most herbaceous annual crops like beans or melons, and two to six times more than corn or sugarcane. The efficiency with which a mezcal plant drinks water to gain edible or drinkable biomass is higher than nearly any other food or beverage crop in the world.

※

During a drought, soil dries out much faster than do the agave's internal tissues. The plants simply slough off most of their root hairs and close off their stomatal "breathing" pores for longer periods of time. The density of those pores on agave leaves is much lower than those of most other plants, so the wax-covered leaves lack the wherewithal to "sweat" out moisture during the dry periods of the year. Whenever drought deepens, agaves retire to a state much like the summertime hibernation or "estivation" of desert tortoises. In this state of "suspended animation," their rates of moisture loss through transpiration trickle down to a bare minimum.

How century plants pull off this trick of conserving water is a knack that some twenty thousand other species of succulents in forty plant families—from aloes to widow's thrills—can finesse to some extent. But very few of these water conservation awardees have been domesticated and grown as commercial crops: A dozen agave species, two kinds of dragon fruit, six columnar cacti, a half dozen prickly pears, one pineapple, and one species of vanilla are the only crops that have this capacity for water frugality.

Each of these plants have independently evolved similar ways to conserve water while producing abundant food and drink. Succulents do so by virtue of a biochemical and physiological sleight of hand called the crassulacean acid metabolism (CAM). That's a lot of syllables to simply say "(We) save water."

Being a CAM succulent boils down to using the most efficient photosynthetic pathway available to any plant on the planet. Agaves can produce an abundance of edible biomass on meager amounts of moisture compared to nearly any tree, shrub, or cereal grain. They stay so hydrated, plump, succulent, and turgid that they often make corn, beans, and squash in the same *milpa* look like pale, wilting wallflowers.

By opening its stomatal pores to take in carbon dioxide during the cooler hours of the night instead of during the heat of the day, an agave hardly lets any water leak from its pores. When agaves do face the searing sun head-on, a special protein called mayahuelin (yes, named for the agave goddess) buffers them from intense temperature shocks that few other plants can tolerate.

That may make agaves the only plants with a protein named to honor a goddess with many breasts for suckling her many children!

With all these assets, agaves are then poised to do their other physiological work of converting sunlight into plant calories during the day when their stomatal pores are closed. While using their photosynthetic machinery during two daily shifts slows down the growth rates of agaves, this double shift allows them to excel in other ways.

Miraculously, they become the slow but steady botanical tortoise that always wins the Sun Belt marathon.

A harvest of agave plants from a single hectare of arable land—about two and a half acres—can yield 25 to 40 metric tons of edible biomass over a ten-year period. Whether converted to food or drink, that is far more bang for the buck for the water invested in its growth than nearly any other crop in the world can muster. On a combined water budget of rainfall and minimal supplemental irrigation of just 20 inches (500 millimeters) per hectare per year, *Agave americana* can produce eight times the dry biomass of corn or cotton in the deserts of Arizona.

As we hinted earlier, the patient growth of an agave far exceeds the cumulative productivity of ten separate harvests of annual crops spread over the same amount of time. The speed demon growth of corn or sugarcane cannot hold a candle to the harvest of a single century plant.

⁂

Although we know that agave has been a staple source of nutrition for millennia, it is difficult to discern just how long ago the farming of agaves began in Mesoamerica and Arid America. We do know that the prehistoric cultivation of a dozen species of agave extended as least as far south as Guatemala and as far north as Arizona, but we still don't know exactly when or where the first domestication of maguey occurred in the 2,000-mile arc of ancient agricultural enterprise. Some have guessed that Oaxaca and Puebla served as the cradle of agave agriculture, while others conjecture that Colima and Jalisco, or San Luis Potosi, Hidalgo, and Querétaro were where agaves were first domesticated. For several reasons, we can't yet figure out where the agaves were originally cultivated on a large scale.

That is because it remains difficult to discern where agaves first passed from the wild to casual cultivation to full-out, formal domestication. Such a genetic transition involved a few of their heritable traits, shaped by the keen eyes of the caretaker and harvester and the sensitive taste buds of the *maestro mezcalero*. It may have also involved increasingly sophisticated techniques for planting, protecting, harvesting, and roasting them.

It seems that the archaeologists' difficulty in dating the onset of agave domestication is the result of two dilemmas.

First, even the best archaeobotanist cannot easily determine whether the agave fibers in a dry cave were all wild harvested from natural habitats, or whether some show signs of favorable responses to cultivation and cultural selection for superior qualities. With little more remaining of the agave leaf than its terminal spine of serrated lateral edge, the field scientist can hardly identify the species that was harvested. It is harder still to definitively confirm that it was from a named variety of domesticated agaves still grown today, as archaeologists can routinely do for cultivars of chile peppers or pumpkins.

In all agaves, the morphological transition from being fully wild to becoming a domesticated plant is not as dramatic as the one between wild teosinte and cultivated corn. The domesticated agaves often grow larger, mature faster, and produce less by seed and more by bulbils and offshoots.

To make matters worse, even when ancient agave remains are found surrounded by the tools of farming and features of agricultural landscapes such as stone terraces, we still cannot be sure that they were exclusively cultivated for alcoholic beverages. It is just as likely that they were grown as *ixtle* or leaf fiber for basketry, as food in the form of fresh flowers or roasted agave hearts or flower stalks, or as feedstock for mildly fermenting beverages such as pulque.

Most conservative archaeologists dismiss the possibility that mezcal distillation was a driver of the first agave domestications. Instead, they argue that prehistoric agave farming began with increased demand for the aguamiel "honey water" from giant magueys to ferment into mildly alcoholic pulque. Until recently, the cultivation of agaves to produce any distilled spirits like mezcal was regarded only as a post-Conquest phenomenon of "mestizo origin."

In essence, historians claim that not until 1565, when Filipino and Chinese traders and their slaves brought Asian stills across the Pacific on the Manila galleons, was there any need to grow thousands of hectares for agaves to produce distilled spirits. Pulque—not the many mezcals, and certainly not tequila—was the prehistoric powerhouse driving agave agriculture.

That is where the great historical sleuthing of Guadalajara lawyer, philanthropist, and agave historian Miguel Claudio Jiménez Vizcarra comes in. He found remarkable evidence of early mezcal production—not just pulque production—in a 1541 report from Tlaltenango in the Mexican province of Nueva Galicia. There, Indigenous residents were drinking jars of homemade "vino" well before grape production for wine-making was established anywhere near there—not even fifty years after the arrival of Columbus in the New World.

Claudio Jiménez is sure that agaves were being made into cognac-like spirits twenty-nine years *before* the first Manila galleon could have conceivably carried to western Mexico the first Filipino or Chinese stills for distilling the juices of coconut or sugarcane. He therefore argues for pre-Filipino distillation in Mexico, if not pre-Columbian agave distillation.

By 1547, Spanish officials were already concerned that many Indigenous inhabitants of the region were getting "plenty drunk" on a vino made from maguey that they fabricated at sites where agaves were being grown in large numbers.

The term "vino" was not only used for wine from fermented grapes at that time. As Miguel and Macario Partida have reminded us, agave plants were traditionally called "mezcal," but the distillates derived from them were simply called "vino" in common parlance, or "vino mezcal" when talking to strangers.

By 1576, the writings of Santiago del Riego—an envoy of new Spanish King Philip II—demonstrate that Nueva Galicians in many towns were already drinking "vino mezcal," a name they have used ever since for the distilled juices of the cultivated agaves. This suggests that they first ushered in the production of wine from native plants, and perhaps distilled spirits as well.

Not only were Mesoamericans already drinking lots of mezcal, but it was coming from their *magueyales,* or tended patches of maguey that had been purposefully planted, and was already being produced on a commercial scale. Its consumption was so common that Santiago del Riego wanted its production to be prohibited.

So, what might the traditional plantings of mezcal have looked like before more European styles of agriculture came along to change them? First and foremost, they were definitely not monocultures of just one kind of agave grown separate from other food and beverage crops such as fruit-bearing prickly pear and columnar cacti, pod-bearing *mezquite* and *guamúchil* trees, intercropped with swaths of maize, amaranths, beans, greens, and squashes between the widely spaced rows of perennial plants.

Even today, from Jalisco to Oaxaca, many campesinos plant and harvest agaves for mezcal *en milpa. Milpa* is their term of choice for a diverse agricultural landscape of mixed annuals and perennials, not limited to corn or its companion beans and squash. But if you were to soar over Mexico and the US Southwest hundreds of years ago, you would see as many terraced *milpas* on hillsides as ones nestled down in the valleys. Mesoamerica and Arid America rivaled Peru and the Philippines for the total land area placed into stair-step terraces. Agaves danced up and down those stairways with the ease of Shirley Temple and Mr. Bojangles, or Fred Astaire and Ginger Rogers.

There are two such traditionally tiered landscape designs, or *trazos,* where agaves remain common fixtures even to this day. Both involve rainwater harvesting and soil retention strategies that sustain the soil moisture and fertility. In this manner, fields of agaves can thrive for decades if not for centuries without depleting the soil tilth, a quality that assures that the sequestration of carbon, water, and nutrients is not squandered. Both agricultural landscape types that generate agaves and tilth were named by the Aztecs long ago, but their use extends well beyond the Nahuatl-speaking agricultural communities in the heart of Mesoamerica.

Metepantle is the first kind of terraced landscape exquisitely suited to producing agaves for mezcal. Our friend, pulque producer and linguist Juan Olmedo, explains that this term comes from the Nahuatl words *metl,* "agave," *tepe,* "hill or ridge," and *apantli,* "line, canal, wall, or terrace." A related term, *nepantla,* refers to the cultivated space or bed within each terrace edged by agaves. Here is where soil—which would otherwise be carried downhill in thunderstorms—is retained, and where a volume of moisture many times that which falls on a rain-fed agave is absorbed by the soils and roots.

Metepantles are widely spaced rows of agaves that form and retain the edge of the dirt terraces that run along the contour of a ridge or hill. Each cultivated row of agaves can have other crops planted above or below them on the terraced slope, such as prickly pear and pitaya cactus, or *mezquite,* *guamúchil,* and *bonete* trees, all of which bear useful products for food and drink. In the *metepantle* system, there are no stone walls edging the terraces, just a web of agave roots to retain the soil. Over time, a terrace-like topography will form. The *metepantle* system was less labor-intensive in construction and maintenance than the stone-lined terraces of *trincheras* or *tepantles* and was therefore more common.

In some parts of Mexico, the terms *trincheras* and *huertas* are also used as names for these food-producing landscapes. In these agroecological systems, slopes are protected from erosion by creating flat terraces that terminate in a cobble walls that structurally secure their rims. *Tepantles* are generally used for growing maize and other herbaceous annual crops in association with agaves. Agave roots reinforce the stability of the terrace wall by holding back even more soil and mycorrhizal networks, while the herbaceous crops leave organic matter and nitrogen to sustain fertility.

Now here is where all that behind-the-scenes physiological magic gets palpable . . . and edible. One way that agaves have found to hang on to the water that their roots absorb is by storing them in special gel-like carbohydrates called *inulins*. Inulins are hygroscopic, which is to say they bind water molecules tightly within their storage molecules so that moisture is lost at a relatively slow rate, even on the hottest, driest days. They not only protect the plant against dehydration, but they reduce the damage caused by catastrophic freezes as well.

Remarkably, the very same qualities that assure that inulins slowly release their moisture from the tissues of a desert agave also serve to slow the digestion and absorption of sugars in roasted agaves when consumed by humans. The special inulins in most agaves roasted to make mezcal— recently named *agavulins*—qualify agave hearts as a "slow-release" or hypoglycemic food. What that means is this: They can help prevent or reduce the impacts of adult-onset diabetes on human health.

There is something wondrous about the way agaves survive challenging conditions. That wondrousness spills over into a glass of mezcal, whenever a slow-maturing agave is harvested, roasted, fermented, then distilled for human consumption.

※

If we learn to think and drink like a century plant does in nature, we too may be better able to adapt to all the challenges that future generations of humans may face in a hotter, drier world. Climate change is upon us, and we need allies (like agaves) that can pull down carbon to reduce the impact of greenhouse gas emissions. We also need to grow food that can take the heat and uses less water than conventional crops. But we also need to learn metaphorically about how to live frugally from our agave allies. Each time we savor agave spirits, we are taking into our own bodies some of the same chemistry of resilience that has allowed them to endure for centuries.

That is right: A bottle of mezcal is imbued with fragrances and flavors that come from agaves sweating their way to survival. It is also loaded with lessons about how to live well on less. Curiously, some of the most powerful chemicals that agaves "sweat out" over their lifetime end up acting as cures for many of humankind's ills as well: terpenes, esters, fatty acids, and saponins. Even before *mezcaleros* cut and roast them, agaves are giving off

Agave diversity below Nevado de Colima's volcanic shield

so many floral, herbal, citric, phenolic, and vanilla-like aromas that they can make your head spin. What's more, agaves give us many hints about how to live well on less.

The hypnotic tastes, euphoric sensations, and wild energy that mezcals offer us simply cannot be matched by any liquor distilled from pampered, overly domesticated wheat, sugarcane, barley, or corn grown in highly manicured fields. Compared to hooch conjured up from these cereal crops, mezcals carry a more enigmatic, fiery spirit. They are a bird of a different feather, one that flies higher, longer, and farther than all the rest.

Now, let us head out into the fields and distilleries, and join in the ceremonies and celebrations to see what we can sip and surmise.

Chapter Three

✺

BIODIVERSITY IN A BOTTLE

Counting species harbored in a bottle of mezcal is far more interesting than counting sheep, because you may get to taste them as soon as the count is over, and you will probably sleep deeply after a nightcap.

Also, it's rare to count very high. In all our years, the most we have ever encountered is an astonishing eleven, from Don Lorenzo Virgen in southern Jalisco. Still, five agaves in a *maestro*'s blend is remarkable, and when we heard that Lorenzo's friend—eighty-year-old Don Macario Partida—had a batch with eight wild and cultivated varieties, we were floored.

Stunned, our ears tried to take in what we had just heard from Miguel Partida—the son of Macario, a master distiller in Zapotitlán de Vadillo, Jalisco. We had already read of the Partida family's vast inventory of agaves recorded by several of our ethnobotanist friends. In 2009, they had identified sixteen cultivated varieties of two agave species in the Partida plantation. That plantation fans out across a ridge above their Chacolo distillery, looking out over the flanks of the Sierra de Manantlán Biosphere Reserve to the west, 125 miles (200 kilometers) from the Pacific coast.

The other side of the Partida property looks out toward the Nevado de Colima volcanic shield, which is assumed to be at the center of early agave distillation in western Mesoamerica. There, in an old pueblo of farmers with

Indigenous Zapotecan ancestry, many of the 6,500 inhabitants of Zapotitlán are engaged with the local diversity of wild agaves, with no fewer than seven native species growing nearby.

By all accounts, the Partida family has made an exceptional commitment to keeping their agave plantings as diverse as the wildlands around them. This commitment delights both mezcal aficionados, who love their novel blends, and conservationists, who are grateful for the many agave stalks they let flower for bats, hummingbirds, and bees.

Nevertheless, until we arrived at the Partida plantation, we had no idea that Don Macario was still recruiting and naming additional, distinctive varieties he had found in the wild barrancas below the Sierra de Manantlán. His son Miguel told us that his father does so to "play around" and to generate completely new blends of tastes and textures in his beverages.

"Now, imagine this," Miguel explained, with an ironic grin on his face, "even though we now have around twenty-five distinct varieties of wild and cultivated agaves we use for distillation, we can't legally call our spirits *mezcal*! That is because of the constraints imposed upon distillers by the way the Regulatory Council for Mezcal interprets its marching orders. And so, we are now simply calling our products *100 percent destilados de agave*, because we are prohibited from using the word *mezcal* on our labels as my neighbors have done for decades!"

Miguel spares nothing when commenting on the bureaucrats who place such obstacles before them: "They've made it far harder for us to sell our mezcal than it is for a farmer to sell dry beans. But why, when we generate so much good? If you look at what we are doing, we are not only conserving the biodiversity of agaves in this Tuxca region. We are also conserving our family's traditions, the *milpa*, the trees, the whole landscape, and our culture as well by the way we do our work. But the policymakers and much of the unacquainted public at large hardly value our work."

The kinds of obstacles that frustrate the Partidas and other *mezcaleros* are counterproductive regulations that undermine the very quality and legacy of their liquors. They place antiseptic constraints over diverse microbial fermentations, and exclusion of "off-types" over using a plethora of species and varieties in one frothing vat, as has been done for centuries. There are also inexorable economic pressures on farmers to "clean up" their "messy" farming practices and mixed-crop *milpas* of all plants other than the single emblematic agave variety.

Despite these legal roadblocks, Miguel, his brothers, and their father Don Macario Partida are having great fun and gaining considerable fame for combining such a diversity of agave fragrances and flavors into a single bottle. For them, and for many other traditional producers who enjoy the myriad flavors of agaves that are part of their cultural patrimony in Meso-america, they are unwilling to sacrifice this diversity.

Says Miguel: "We are resisting all of this. We love the work we do, but poverty is never too far away from our door. . . . We're grateful that we are now getting more respect for what we do from some of you, but the way we use so many agaves should have already given us a place at the top of the spirits market. That hasn't happened yet."

❧

If variety is the spice of life, then let it be known that Oaxaca dwarfs any other state or province in the spiciness and splendor of its mezcals. At least forty-three wild and several more semi-cultivated species of agaves flour-ish in agrarian Oaxacan landscapes, though many are shared with Mixtec *mezcaleros* in the southern reaches of the adjacent state of Puebla. There are undoubtedly more to be discovered, for they may be hidden away in the most remote reaches of that southern state. Remarkably, a good many of these agaves make their way into mezcals or other Indigenous products of Oaxaca, from ropes and basketry to pulque and edible flowers on your dinner plate. Some 600 miles (965 kilometers) southeast of the Partida's distillery, we find the cradle of agave diversity on the Puebla-Oaxaca border.

Driving southeastward from the town of Tehuacán, Puebla, toward Cuicatlán, Oaxaca, we find ourselves staring out the window at more per-plexing shapes, sizes, and splashy colors of wild agaves and columnar cacti than can be seen anywhere else on Earth. The hot, dry Tehuacán-Cuicatlán Valley is Mexico's geographic showcase of wild agave diversity, sometimes with twenty wild agave species featured in the same habitat.

The valley is at the heart of the region where wild agaves began to evolve and diversify some ten million years ago. They are sprawling on cliff faces and edging the rocky banks along watercourses amidst the richness of 2,700 distinct kinds of plants that occur in that arid valley, with a third of them nowhere else on the planet.

Over the last two decades, botanists have described five new species of agaves every four years, many of them from Oaxaca. When Howard Gentry published his *Agaves of Continental North America* in 1982, he identified just 136 unique species. Today, of the 215 described and widely recognized, three-quarters of them are from Mexico. When other recently discovered agaves are fully described in print, perhaps as many as 250 of their species will be known to science. The genus of *Agave* now stands tall among the three largest groups (or genera) of wild plants in all of Mexico.

National University (UNAM) professors Luis Eguiarte and Valeria Souza Saldívar have remarked that since the agave emerged as a distinctive genus or taxonomic grouping of plants ten million years ago, it has undergone one of the most spectacular diversifications of any plants in the Americas. We have already introduced to you the term *adaptive radiation* to describe how a group of plants fans out to live in a variety of habitats. Agaves do this better than any other group of "monocot" plants on the planet, outdistancing all grasses, sedges, palms, tulips, irises, onions, bananas, and gingers. There may be as many as forty-two species of agaves currently used to produce mezcals in twenty-four Mexican states, and well over two hundred folk varieties recognized by *mezcaleros* themselves.

⚶

Numbers alone may numb your mind, but the diversity of flavors and fragrances that these agaves carry in their hearts may perk you up to pleasures and wonders that cannot be experienced anywhere else.

Consider how different the aromas and taste notes might be between just two of the many agave species used for mezcal in Oaxaca. The two Oaxacan mainstays are *Agave angustifolia* (which includes *espadín*, *espadilla*, and *zapupe*) and *Agave potatorum* (which includes *tobalá* and *papalometl*). When Araceli Vera-Guzmán and her colleagues at the Oaxaca Center for Integrated Regional Development compared the fragrances from just these two

Oaxacan mezcals, they were amazed by what was revealed. The two species differed by at least twenty-six volatile compounds that influence both what we smell and what (we think) we taste. Since four-fifths of the "flavors" we claim to taste are aromas that hit our nose before they hit our taste buds, having a variety of volatiles is essential to a memorably aromatic mezcal.

What is more, distinctive profiles of volatiles in Oaxacan mezcals are attributed to their genetic traits more than to the environmental factors found where they were grown. Their wild genetics are more important than any other factor in determining their differences in aromas. Of course, many human factors also shape their profiles of fragrance and flavor: how they are processed, what woods are used at the distillery, and the many microbiological contributions of yeasts and bacteria that do most of the work to ferment them.

If we extrapolate from the two kinds of Oaxacan agaves that Dr. Vera-Guzmán selected for her study to embrace the thirty-three major agave species commonly used for making mezcals across Mexico, our "sensory" or "olfactory" world expands exponentially.

Let us put this into a global perspective. Compared to the single cultivar allowed in tequila, there remains an enormous diversity of wild agave *species* utilized in the making of mezcals. This gives mezcal production much more flexibility in their environmental tolerances, and a wider range of tastes and aromas.

There are twenty-four species or varieties of other agaves that have been domesticated from the edges of Central America clear to the Grand Canyon. Six of these domesticated agaves were recently described by our friend Wendy Hodgson in Arizona, where they were grown prehistorically near the northern limits of the natural range of agave genus. In fact, the number of agave domesticates from the Arid American region of northwestern Mexico and the US Southwest now rivals the number domesticated in any region of Mesoamerica, which stretches from the Tropic of Cancer near Durango and Mazatlán, Mexico, clear to Central America.

Keep in mind that just one species of wild maize was ever domesticated *(Zea mays)*, and it is used in all corn whiskeys and moonshines. As with maize, there is just one cane crop found in the materia prima of rums; a single, genetically homogeneous clonal species of sugarcane is employed in all of them.

❋

That said, some other spirits are inching in the right direction. Rye whiskey must be made of 51 percent grains from rye plus one or two other grain species, usually barley or maize. And three grain species is the limit on what you can find in bourbon as well, which is usually a mix of corn, malted barley, and rye.

Today, many vodkas still use fermented potatoes, but wheat, rye, rice, sorghum, or maize are acceptable in vodkas. In Japan, *sochu* is usually made with just one species of short-grained japonica rice, or occasionally from barley, sweet potatoes, or buckwheat. Gins are richer, for they can draw on four or five different grains, but also include a dozen different infused botanicals, which are typically added in minuscule quantities after the first distillation.

Drink any other major spirit distilled on this planet, and you are usually imbibing one to three different species of grains or root crops. With mezcals, you may be blessed by as many as ten different wild agave varieties in a single bottle, or a dozen distinctive agave species distilled in a single state, such as Oaxaca.

❋

When some of our gin-loving colleagues hear that declaration from us, they counter by claiming that the use of a dozen or so botanicals as natural flavor enhancers in artisanal gins rivals the plant diversity found in mezcals. But if we start counting all the herbal flavors used to infuse gins, let us also count the many herbs, fruits, and game animals used to infuse mezcals as *curados*, or those special cases known as *pechugas*. When it comes to the richness of herbs infused, gins are far outpaced by mezcals.

Pechugas are mezcals that are made by hanging plant and animal products in bags or on strings below the condensers of certain stills, so that their aromas are carried up with the vapors arising from the agave must and condense into the alcohols being distilled. Making *pechugas* is just one of many mezcal traditions for enhancing agave spirits with herbal, floral, and faunal infusions.

When we count the animals now used in making *pechuga* mezcals—such as breast meat from turkeys and laying hens, hams from Iberian hogs, legs from lambs, rabbits, chickens, quail, and deer, or tails from iguanas and rattlesnakes—we can add ten sets of animal products that give their aromas to our list of what is embedded in mezcal diversity. For as much sleuthing as we've been able to, we simply can't come upon the use of *any* animal products in other distillates of international renown, except Himalayan deer musk grains in medicinal tinctures. But when we toss in the entire cornucopia of wild fruits, roots, barks, herbs, grains, and flowers added to *pechuga* mezcals or used to infuse *reposados*, we feel like we are trying to tally all the species that hopped aboard Noah's ark!

Who would have thought that *capulín* sand cherries, *uvalama* berries, *tuna* cactus fruit, *bellota* acorns, and Mexican tarragons known as *yerbanis* would be hiding in your bottle of mezcal? How would you know that spices like anise, cinnamon, and ginger, or grasses like rice, criolla corn, and *te de limón* would be among the thirty species of plants other than agaves themselves that pour out into your shot glass or *jicarita*? Dozens of the "botanicals" and "zoologicals" in mezcal were historically borrowed from the ancient Mesoamerican diet, and some of them have been used to flavor *pulques* fermented from the aguamiel of giant agaves for millennia.

※

The infamous "red worms"—*gusanos rojos* or *chilocuiles*—are really larvae that have been drowning in the bottom of bottles of mezcal since 1950. That is when artist Jacobo Lozano Páez first reckoned that they would be a good gimmick for marketing Oaxacan mezcals.

The red larvae and the ground-up powders of two moths, *Hypopta agavis* and *Comadia redtenbacheri*, are sometimes roasted, dried, and added to salts with hot peppers to be consumed with agave distillates. Some *mezcaleros* favor the *gusanos blancos* or *meocuiles* that are the creamy-white larvae of *Aegiale hesperiaris*, the tequila giant skipper butterfly, which is a pest on *tequilana azul* and other agave plants.

There are many other kinds of larvae, grasshoppers, and beetles that are now being ground into powder and added as flavorants to accompany mezcals, either mixed with salts or as standalones, including the sisal or agave snout weevil, *Scyphophorus acupuntatus*. While the emphasis on these additions has been on their supposed impact on flavor, these infusions originally had medicinal and cultural purposes.

Rather than the 2-inch worm many consumers have come to expect, Michoacán *maestro mezcalero* Jorge Pérez sometimes infuses his batches with a whole rattlesnake. The serpent has little impact on flavor, but the infusion is seen as an appropriate topical remedy for rheumatism and muscle pain. On the next shelf in Don Jorge's *palenque*, one may find a large jug of mezcal with tiny white pellets piled up in the bottom third—"Mexican caviar" or, for the uninitiated, ant eggs.

Surprisingly, these animals do not always impact flavor when added directly to the distillate. Dried and ground, larvae and weevils can certainly offer an earthy, musty, umami-rich flavor to salt mixtures, while insects like *chapulín* grasshoppers offer a savory but nuanced and nutty fragrance. Mixed with a little hot pepper, these salts can make for an appealing rim on a cocktail glass, but outside of homeopathic medicine, that's as far as the larvae have an impact.

There has been a long-maintained myth about mezcal: that a worm is some marker of quality or authenticity. In fact, mezcals with worms are typically low-quality commercial distillates. Agaves with worms are diseased agaves, so you might wish that the worm came from a bait 'n' tackle shop and not anywhere near the fields that yielded the mezcal.

The lingering debate about the influence of larvae and grasshoppers on the flavor of mezcal itself may have prompted the agave industry and the public to stand up against some rather silly regulations. They expressed indignant outrage when Mexican government agencies attempted to eliminate all worms and weevils from their distillates in 2005. Some bureaucrat thought that promoting wormless, "antiseptic" mezcals would be a good way to bolster foreign consumer confidence in the quality of agave distillates exported from their homeland.

But as *maestra mezcalera* Graciela Ángeles Carreño of Santa Catarina

Minas lamented to news reporters at the time, the wormy ship had already sailed out of the harbor: "If there is no worm, there will be no sales."

✳

Even if it remains a bit misunderstood by most mainstream consumers, the biodiversity in a bottle of mezcal remains extraordinary. We cannot round out our celebration without also noting the trees that influence the aromas of certain mezcal. Let us remember the many kinds of trees used as firewood in roasting pits and beneath the stills themselves, or used in the fermentation vats or barrels where aged mezcals settle on their angle of repose.

The *mezcaleros* we know are very fussy about their choice of woods for roasting so as to enhance or mute the smoky fragrances carried into the bottles. Some prefer *encino* (oak); others *mezquite*; still others *guamúchil* or *sabino*. And now, because of the ever-higher numbers of trees that have been cut from wildland forests and savannas to fuel the mezcal industry, *mezcalera* Sósima Olivera is insisting that her Tres Colibrí Cooperative grow their own tree crops of favored species to reduce the environmental impacts of their Mezcal FaneKantsini.

Fermentation vats in Mexico are built from local woods, often *parota, paine,* and *oyamel,* while oak barrels for aging agave spirits in repose usually come used from whiskey and bourbon distilleries in the American Midwest and cognac houses in southern Europe. Each wooden barrel imparts a unique flavor of its own.

Because wood is more porous than clay, clay packs in the natural flavors of pit-roasted agave, whereas wood may allow for more circulation of fragrances in the batch. It fosters a molecular exchange that alters the end product, specific to the type and source of the wood. In other words, wood brings a unique expression of fragrance and flavor to the terroir of the spirit.

Moreover, the wood's gifts to the complexity of agave spirits—even when the stills have been hidden in the hills by bootleggers—live and breathe forever in the spirits we enjoy. The choice of wood for a still is one of many influences on the flavor profile of a mezcal, but it brings the landscape signature of desert, savanna, selva, barranca, or sierra right into the spirits in our cup.

While worms, labels, and billboards often take the spotlight, it is these natural and less visible elements that offer the sense of time and place that we
treasure in these spirits. Of course, most of the firewood, game, herbs, roots,
and fruits contributing to the richness of traditional mezcals have been harvested from the wild. The pollinators of agaves—three nectar-feeding bats,
over a dozen hummingbirds, several dozens of species of bees, and a googol
of hawkmoths—are almost all wildlings as well.

Together, all these wild species impart to mezcal unbridled tastes and
aromas that seldom get across the gate into the farms where toned-down
domesticated crops are grown for other spirits. They carry into the distillate secondary chemicals such as volatile oils that are at the heart of all fragrances and flavors in wine, gin, herbs, and most vegetables as well.

Let's just put it this way: Artisanal mezcals are so nuanced, so awe-inspir-
ing, and so memorable because they carry along with them the flavors and
fragrances from a chorus involving hundreds of species of microbes, insects,
and other wild lives, each of which adds its own special note to the song.

And yet that diversity has never been adequately safeguarded. Even the rare
agaves that are federally listed as threatened or endangered by Mexico's
SEMARNAT agency or by the international "red-list" of CITES are not sufficiently protected in their habitats due to a lack of sufficient support for the
available well-trained caretakers or park guards. In Jalisco, there is concern
about declines due to the overexploitation of threatened agaves for *raicilla*
production, and due to the disease transmission to wild populations from
tequila plantations.

At the same time, dramatic reduction in the population sizes of wild agave
patches is also occurring for many patches that are not legally protected by
endangered species and protected area regulations. From Oaxaca northward
to within a mile of the Arizona border with Sonora, independent reports
have been coming in from at least six states that wild agaves of several species are suffering from clandestine *saqueo*, or removal from their habitats.

These clandestine thefts may be occurring in response to critical shortages of both wild and domesticated agaves, depending on the locality and species. We have heard unconfirmed reports that in Oaxaca nocturnal harvesters are digging up wild agaves planted just months before in reforestation efforts sponsored by governments, non-profits, and distilleries themselves.

We have heard frequent reports of marked declines in population of wild species, even in remote reaches of the Sierra Madre Occidental, where both overharvesting and climate disasters have diminished patch size. Some of the pressure is driven by the current high demand for mezcals using both wild and cultivated agaves. By 2022, mezcal production had reached more than 8 million liters a year. It is hard to imagine how the slow-growing species of agave can keep pace with such demand for the many mezcals, let alone for *raicilla* or *bacanora*.

Meanwhile, dramatic shortages of cultivated plantlets of *tequilana azul* from nursery stock have been reported. From 1878 until 1968, the volume of tequila annually produced in Jalisco never topped 20 million liters, when cultivated agaves of eight varieties were harvested from fields. By 2022, over 534 million liters of tequila were produced in Mexico in good years— 26 times the usual volume that was produced before 1968. Because there are relatively few high-volume nurseries in Tequila country, just a few brokers influence both the price and volume that are presented to potential buyers each year.

As we shall explore in greater detail in subsequent chapters, neither the tissue culture labs nor field nurseries for *tequilana azul* have been able to keep pace with the still-rising demand for new agave plants in recent years. In some years, it has been reported that planters fall tens of millions of plants short of what they need to replace the number of agaves they had just harvested. As demand outstripped supply of *tequilana azul* "seeds" or young transplants for distillation, the price per kilo of this propagation material tripled in a matter of years. But now there is also competing demands on this materia prima for agave nectars, landscaping, and biofuels.

If the nurseries associated with the largest and most lucrative tequila distilleries in Mexico are already failing to keep up with the demand, how will the smaller, more artisanal producers fare? And how will the smaller, often more rustic nurseries found sporadically in mezcal-producing states ever keep up with the exponential growth curve for mezcal sales?

Master tequila distiller Carlos Camarena predicted to Jack Robertiello of *Beverage Journal* that this will soon be recognized as the most severe agave shortage of domesticated varieties like *tequilana azul* and *espadín* ever. He said, "In the past, the shortages only lasted a couple of years and then we'd have another surplus. Right now, we are in about the third year of shortages and I expect this to last at least another couple years before things balance out."

Even the Mezcal Regulatory Counsel (CRM) was honest enough to admit that the current annual demand for 1.3 million gallons (5 million liters) of mezcal worldwide was negatively affecting the availability of plants belonging to fifty-eight kinds of wild and cultivated varieties. As Nacho Torres and the staff of Milpa A.C. revealed from data in the CRM's own reports, "Clearly, the evolution and increase of mezcal activity over the past decade has had a profound impact on the environment, when we see that eighteen of the fifty-eight total varieties—31 percent of the total—are now in the list of the most vulnerable, close to extinction."

Nearly a third?! Such a drastic decline in diversity is hard to analogize without sounding alarmist—a third of music genres becoming unavailable is indeed hard to imagine—so we offer conceptualizations based on some of agave's many uses.

How tall could our buildings be without a third of the construction materials they regularly require? How well could we and our neighbors live without a third of our medicines? Such stresses on supply chains are almost too disruptive to even imagine. Or what if we lost a third of the plant species in our food supply in one fell swoop? It seems unfathomable, at least for now.

The report goes on to indicate that nearly 7 percent of all mezcal production from 2019 to 2022 came from three wild agave varieties, *Agave potatorum, Agave karwinskii*, and *Agave cupreata,* and that those three species are now among the most vulnerable in Mexico.

Recently, a brilliant team of field scientists from the National University's Institute of Biology estimated that these wild species can support no more than

10 to 30 percent of the mature plants being harvested each year if they are to maintain viable populations for sustained production. And at this point, there are not enough agaves being propagated in reforestation nurseries to offset the harvests of wild agaves. Further, the report indicates that "an increasing demand for single-variety distillates has also put in danger other species, such as *Agave marmorata*, 'tepextate,' which is nearly threatened enough to be included in the Mexican list of wild species considered to be vulnerable to extinction."

The cold, hard facts suggest that half the wild species used in some Mexican states are being extracted to make agave spirits, but with drought, wildfires, and increased harvesting pressure since the 2015 mezcal boom, the supplies may not be able to keep up with the demand. For mezcal lovers, the decline or local extirpation of these agaves would be a terrible loss. The lovers of these spirits do not want to imagine losing a third of the diversity of their favorite mezcals. And yet, according to our conservationist colleague Nacho Torres, the loss of so many agaves as viable populations may be a biodiversity tragedy even worse than the loss of high-profile jaguars, a single species.

It may come as a surprise to drinkers that we are losing mezcals, given the rising availability of these spirits all around the world. We were amazed in 2019, when the IWSR—a leading source of intelligence on the alcoholic beverage market—announced that mezcals' compound average growth rate (25 percent) had outstripped expectations over the past five years and was four times that of tequila's growth over the same period. By 2019, the combined production of all mezcals and 100 percent agave distillates had approached 2 million gallons (7.5 million liters) in volume, and that made many mezcal lovers ecstatic.

Twenty months later, a Future Market Insights report projected that mezcal sales would reach $733 million by 2027, forecasting a lower but steadier growth rate of near 14 percent for the period between 2020 and 2027. What's more, the mezcal boom that began in 1995—with domesticated varieties like *espadín* taking the lion's share of the market—had since diversified, embracing many varietals of wild agaves in addition to *ensambles*, or blends of wild and semicultivated mezcal varieties as well.

❋

And yet that "good news" was followed by other reports that gave us pause, if not paroxysms of panic. When journalist John Kell asked the spirits world, "Has mezcal become too big for its own good"? in the pages of *Fortune* magazine, many starry-eyed mezcal aficionados finally began to take notice. Then *National Geographic* ran two stories on the downsides of mezcal's rise to fame in just three years. The first drew attention to ways in which the mezcal boom triggered the "dwindling of the agave," while the second argued that the boom was "bad news for bats." How will the emerging but already enormous demand for quality mezcals continue to affect the already-imperiled plants, people, and places that underpin this tradition?

To keep agave production from becoming just one more form of industrialized agriculture, should we harvest only wild agaves from reforested areas, or do wild populations of agaves now need to be safeguarded from overharvesting?

How has demand prompted changes in the predominant practices of fermentation and distillation, positive or negative? Are the humans whose faces and hands we identify with mezcal adequately compensated for their physical work and intelligence, from harvester to *cantinero* bartender? What good news stories of innovation and restoration can we tell that will inspire others to do the right thing? Frankly, these are tough questions to answer with smooth words and easy sound bites.

For now, let us simply concede that the production of mezcal is facing dilemmas as daunting as those being faced by the tequila industry, but the key difference is that tequila comes completely from domesticated plants in cultivated environments, while mezcal still has a wild side to it.

❋

When we compiled the various newspaper articles and technical reports from government and industry on the shortage of agaves, we realized that there is clearly an impending crisis, but that many journalists were "mixing apples and oranges" in their press dispatches. They were often confounding the plant conservation issues of overharvesting slow-growing wild agave species with the industry's shortage of nursery stock of domesticated agave plantlets that are transplanted into agricultural fields.

Although both wild and domesticated agaves are now grown in nurseries, the pressures on these two sources of materia prima for mezcal are very different, even though the scarcity of one of the sources may inevitably affect the other.

To tease apart what is happening with truly wild agaves that form natural populations in the deserts, woodlands, and forests of Mexico, we turned to our friends who investigate wild agaves for the Species Survival Commission, which works under the umbrella of the International Union for World Conservation (IUCN). The IUCN is the global body of scientists, conservational professionals, and policymakers that has published the official Red List of Threatened Species since 1964, using carefully evaluated data compiled by specialist groups on different sets of wild plants and animals.

At a workshop that was part of the 2019 Agave Heritage Festival in Tucson, we tracked down Raul Puente-Martínez, Curator of Living Collections at the Desert Botanical Garden, and one of the lead members of the SSC Specialist Groups helping the IUCN determine how many agaves are threatened and why. What Dr. Puente could offer us as a *preliminary* assessment of the conservation status of agave diversity—and the threats to the agaves used for mezcal—blew us away.

In the historic library at the Desert Laboratory on Tumamoc Hill in downtown Tucson, where agave scientists have been gathering for over a century, Puente was as blunt as a botanist can be: "We have already surveyed the status in the wild of over 180 species of agave, and what we can say is that at least 50 percent of them qualify to be included in the categories of different levels of threat that the IUCN uses to determine which wild plants and animals are truly of conservation concern. Most of these species are in Mexico, of course, since it is the country with the greatest diversity of agaves. And many of them are the species of agave that have long been used for making mezcal or for other uses."

To be sure, there are factors other than use—or overuse—that are affecting the survival of agaves. For starters, 60 percent of all agaves in Mexico have highly localized distributions; they are what floristic geographers call "micro-areal endemics"—plants whose ranges hardly cover a state or two, for they naturally grow in an area smaller than a 360-mile (580-kilometer) square. (Compare that to the broad geographic range of *Agave angustifolia*, the wild species that is the ancestor of the blue tequila agave, which begins just 100 miles [160 kilometers] south of the US-Mexico border and extends south all the way to Nicaragua and Honduras!)

The fact that three-fifths of all agaves are narrowly adapted and restricted to landscapes is one reason why their distilled liquors are so highly prized by local communities. Their "taste of place" or terroir is not transferable to production areas in other states or countries. Their architectural or morphological adaptations to a particular terrain—as well as their biochemical defenses—are elegantly specialized to fit the unique climatic and ecological challenges of growing in one very particular environment.

All this is to say that some agaves would be "rare" whether they were ever harvested for mezcal or for other uses. Yet, rarity—in and of itself—does not mean that they are inevitably at risk of extinction. That said, if an agave already has a small growing range and a few scattered populations within it, factors like climate change and overharvesting can quickly push it toward endangerment. This is what our friend Alejandro Casas revealed in detailed studies with his team of "Agave All-Stars" in the Tehuacán-Cuicatlán Valley.

By nearly any measure, the Tehuacán-Cuicatlán Valley is among the most diverse and singularly stunning arid landscapes in all of Mexico. We were blessed by the chance to travel into the valley with a team of seasoned Mexican scholars led by Alejandro, who has explored plant-culture interactions there over many decades. He took us to one of the very sites where his long-term research has taken place, the Helia Bravo Botanical Garden. It was clear to us that what his team had revealed in its many studies was a harbinger of trends occurring in many other parts of Mesoamerica.

Alejandro charged one of his many brilliant students, América Delgado-Lemus, to figure out what human uses make already-rare agaves more vulnerable to local extirpation—a term that implies loss of one population after another until species extinction occurs, like a row of falling dominos. América went to the poorer villages in one part of the valley, asking residents how each of the agaves present there was used. She found fifteen different traditional uses of agaves in addition to roasting them to ferment and distill mezcal. She also recorded whether their products were exclusively extracted from wild plants in wild habitats, from semi-managed areas of reforestation, or from domesticated plants in *milpas* or other kinds of agricultural settings.

Once she and others on Alejandro's team had interviewed campesinos about how they used twenty-eight out of the thirty-four species that occur in the Tehuacán Valley, Alejandro brought his team together to pore over the data in the hopes of finding patterns that might explain which agaves were at risk and why. The team found that thirteen kinds of Tehuacán agaves were fermented into mildly alcoholic pulque, but only seven of those species were targeted for mezcal distillation in the valley.

Of the seven, five kinds of agaves were most favored by Tehuacano *mezcaleros*. They were exclusively harvested from wild populations out in the dry, subtropical thornscrub, or in other kinds of natural or managed vegetation higher up on the slopes of the valley. And while the fifteen other uses of these agaves hardly affected their population size, harvesting mezcal agaves from the wild inevitably generated a greater impact on the wild population remaining.

Traditionally, harvesters took no more than two hundred plants at a time—typically from the same stand—for a single roasting and made only a few batches of mezcal each year. According to experienced elders in the valley, none of the five agave species preferred for mezcal production were in any way endangered fifty years ago.

It was not until the mezcal boom in the '90s heightened demand that the *mezcaleros* increased the number of pit roastings or *tatemadas* that they underwent each year, each harvesting thousands rather than hundreds of plants annually. Some of them also increased the number of agaves they could roast at one time by excavating and lining larger roasting pits. But being a *mezcalero* shifted from a family pastime on weekends and holidays to a full-time occupation for several family members and their friends.

Sometime between 1974 and 1994, the external (national and international) demand for mezcal began to push many populations of agave past their tipping point. Agave growth and reproduction could no longer keep pace with consumption. As demand surged, agaves in the Tehuacán Valley dwindled.

Population after population of the five rarest agaves were extirpated, "blinking out" like lights that had lost the energy to keep on burning. Alejandro Casas and América Delgado-Lemus were compelled to conclude that the mezcal boom had encouraged Tehuacano families to more intensively harvest agaves. To be sure, mezcal distillation had become the primary (though not the only) human use that pushed agaves toward endangerment.

This decline was not a trend that any of the campesinos wanted to see continue. They asked Alejandro for help in agave propagation to reforest the area with native agave species. Alejandro explained the community-based collaboration in this manner when we were together in Tehuacán in the summer of 2022: "The people of San Luis Atolotitlán had asked us to help them with reforesting the papalometl agave because the populations of this species had slowly been depleted in their area. So we helped them with a greenhouse and nursery, where they established small agaves to transplant out. But when they first began putting them out on the lands, they put them out in the open, in full sun. Ninety percent of the transplants died. We then helped them think through how to use nurse plants—native trees and shrubs that can provide a shaded buffer from heat or cold to protect them. Now, most of the more recent outplantings survive and establish themselves well in the shade of shrubs or trees."

We are in no way trying to disparage or villainize the many conscientious *mezcaleros* of Puebla and adjacent Oaxaca for trying to respond to global demand by upping their production of mezcals. But there are indeed tried-and-true ways to meet the demand for mezcal without depleting populations of agaves or the firewood species that they rely upon to roast and distill their agave distillates.

The intensifying pressures on wild mezcal diversity today in Oaxaca, Puebla, Michoacán, Durango, Jalisco, and Sonora are not merely due to the vagaries of rainfall variations and agave yields from year to year or decade or decade. They are tightly linked to the rising global demand for mezcal. We all need to invest in agave restoration and reforestation in one manner or another, or the well we love to drink from will run dry.

Agave goddess Mayahuel caring for her plants

Chapter Four

✸

MEZCAL AS CULTURE

You may have seen them featured in museums: tawdry objects that archeologists euphemistically term "coprolites" (that is, paleofeces) and others called "quids." Some of the earliest known coprolites from arid reaches of North America are fossilized human feces with obvious chunks of agave fibers embedded in the not-so-appealing matrix of old turds. The quids are the fibrous refuse of roasted agave leaves that were not tossed away after the sweet pulp was scraped off them by human mastication, for they were left in the sun to dry for later use as prehistoric "handy wipes" or "personal hygiene pads." These artifacts are among the oldest evidence of agave use in the New World. They may predate the use of shot glasses and bar stools by thousands of years.

The late great archeologist of prehistoric diets, Vaughn Motley Bryant Jr., not only visited but pulled hundreds of samples from rock shelters and caves above both the Pecos River and Río Grande. The oldest identifiable artifacts in this region that Bryant and his gifted student Kristin Sobolik encountered came from human coprolites containing agave fibers dating back ten thousand years. They confirm that agaves were among the first plants used for food and drink by early desert hunter-gatherers in arid North America. As archaeologists Suzanne and Paul Fish have confirmed, agave was used for everything from bedding material to clothing, sandals, teas, food, weapons, and even as a type of chewing gum. Quids were wadded up agave fibers that were continually masticated and then spit out into a type of tobacco-chaw type wad, many times with teeth indentations still intact.

Of course, many of the coprolites and quids that were amassed in dry caves have carbonized fibers in them, for they were either toasted on the coals of a wood fire or pit-roasted below ground before being chewed over; humans cannot metabolize raw agave. They are irrefutable proof that the fossilized remains are not just raw *biofacts,* but cooked *artifacts*—features of cultural processing and expression. The ancient art of cooking had already led to the inclusion of agaves in the Indigenous cultural traditions and diets of North America.

Sobolik and her collaborator Jeff Leach have elaborated on Bryant's earlier work, taking his hypothesis about the prehistoric importance of these succulents even further. Through applying a series of analytic techniques, they have demonstrated that agave foods and beverages have been central to human nourishment in North American deserts for at least ten millennia.

When analyzing the ratios of plants that use different metabolic pathways found in desert floras, they confirmed that agaves and their close relatives, sotols, were among the dominant components of pre-Columbian diets along the Río Grande. In the *British Journal of Nutrition* in 2010, Leach and Sobolik reported that desert dwellers relied on a greater percentage of succulent plants in their diet than the dominance of fruits from various trees and vines found in the diets of tropical dwellers in the wetter reaches of Mesoamerica.

In fact, dwellers of the Chihuahuan Desert may have consumed one of the highest concentrations of dietary fiber in the form of inulins from digesting agaves of any diet so far analyzed from any region of Planet Desert. These inulin prebiotics stimulate the growth and development of beneficial gut bacteria. They suppress the growth of less desirable or deleterious microorganisms, while at the same time generating a diverse gut biota that promotes health and well-being.

Sobolik and Leach's analyses of prehistoric plant remains and paleofeces indicate that desert-dwelling men were consuming roughly 135 grams of agave inulins on an average day, while women in the same pre-Columbian era were consuming 108 grams. That is about five times the prebiotic fiber that most Americans today are gaining from all plants in their diets.

As Sobilik has concluded, agaves were contributing such a high level of prebiotic and probiotic inulin-type fructans that it is "difficult to comprehend by modern standards. . . . We modern humans try to create that healthy gut environment by eating probiotic products with positive microorganisms added, rather than eating the foods—agaves—that naturally create the foundationally positive gut substrate."

Though many desert fruits, stalks, and roots can be eaten raw, most parts of the agave plant need to be cooked to render their biomass palatable. Among the earliest and most definitive arts that characterize Indigenous American cultures are the arts of smoking and grilling meats and plants (as *barbacoas* or barbecues), as well as the arts associated with pit-roasting or baking carbohydrate-rich plants.

If we delve deep into Indigenous scholar-activist Victor Manuel Rojo Leyva's reflections of the cultural roots of the term *mezcal*, we see that it is not only about etymology, but artistic metaphor: "The word has its roots in the vocabulary of the Nahuatl language, either as *metl* (agave) and *ixcalli* (cooked) that give us the Hispanicized term *mexcalli* (roasted or baked agave); [or] the roots lead back metaphorically to the terms *metzli* (moon) and *calli* (house), to signify that *mezcal* is symbolically 'the home of the moon.'"

The second meaning of the word *mezcal* metaphorically implies transformation of a wild, sometimes caustic plant into a dreamy domain, not just into a delectable beverage. It may also suggest that the curvaceous leaves of many agaves echo the shape of crescent moons huddled together to form a home. This metaphorical leap reminds us that agaves are much more than a utilitarian commodity; as we noted earlier, they are granted an exulted status, one infused with wonder and spirit. They artistically inspire us to look beyond the flesh into the spiritual significance of this "cultured" plant to Indigenous cultures.

Look no further for the cultural significance of agaves than the handful of surviving pre-Columbian codices and at least five hundred colonial-era Aztec manuscripts called *Mēxihcatl* in classic Nahuatl. There, you will find hundreds of glyphs that illustrate agaves in the flesh and the agave goddess Mayahuel in spirit.

Mayahuel was the goddess of nurturing, nourishment, and fecundity. Her four hundred breasts flowed with milky-colored pulque, the nutritious and slightly intoxicating beverage. She suckled her four hundred booze-loving children collectively known as the Centzon Tōtōchtin.

Curiously, it is true that some agaves have four hundred nipple-like "teats"—the jutting protrusions on undulating leaf edges that agave botanist Gentry drolly defined as "fleshy prominences under the teeth on the leaf margins." He counted how many nipple-like spines there were per teat on the lateral margins of various agaves and decided to use this trait as a diagnostic marker of one species as opposed to another.

Images of Mayahuel's voluptuous feminine figure emerging out of a giant agave run rampant in the Codex Borgia and the Codex Borbonicus (or Cihuacōātl), both from the sixteenth century. For Mesoamerican scholars, these priceless pictorial accounts from the earliest days after the Spanish conquest are the cosmological equivalent of the Dead Sea Scrolls or the Gutenberg Bible. Mayahuel the Agave Goddess is also prominently featured in the Codex Ríos, and the Codex Magliabechiano. Her iconic shapeliness was also carved into stone at the Great Pyramid or Templo Mayor of Tenochtitlán, which was first built in the capitol of Nahuatl-speaking Mexican people around 1325 CE to 1440 CE.

Tragically, in one rendition of a Mexican legend, Mayahuel was dismembered, then eaten by demons. When Queztalcóatl realized that he had lost his lover, he planted the remains of her body in the soil, which then reemerged from the earth as plants of maguey.

As an icon of the regenerative and nourishing capacity inherent in all of us, intoxication, and fertility, Mayahuel is so ubiquitous in Aztec imagery that the Mexican scholar Oswaldo Gonçalves de Lima dedicated 278 pages to her iconography. His 1956 classic, *El Maguey y El Pulque en Los Codices Mexicanos*, includes one of the few descriptions of her power over people taken from the Codex Magliabechiano: "The next demon or deity was called Maya-huel, which means the maguey [spirit], because the juices which comes from her were an intoxicant. And so, she is nearly always portrayed dancing."

※

Keep in mind that millions upon millions of Mesoamericans have been nourished by the milky, sweet juices of aguamiel, "agave honey water," that were fermented into pulque (or *octli*) over the thousands of years that agaves were cultivated for food and beverage. The upwelling sap (fresh aguamiel) of giant maguey plants is sucked up into a gourd by their harvesters, then drained into a bowl until there is enough to ferment in bags made of animal skins. It is these skin bags that harbor dozens of wild yeasts and bacteria, which ferment the prebiotic sap into a frothy probiotic beverage called pul-que, which typically runs 4 to 6 percent in alcoholic content.

These two agave beverages—fresh aguamiel and fermented pulque—were as common in central Mexico up until World War II as wines were in Europe, or as hard cider was for Yankees in Frontier America. Babies were weaned off their own mother's milk by the upwelling of aguamiel from the incarnate body of Mayahuel, and later graduated to the mildly intoxicating ferment of pulque . . . and perhaps, later, to the distilled essence of maguey in sacramental sips of mezcal spirits.

Up until the last few decades, the male-dominated cantinas known as *pulquerías* were the primary social and political "watering grounds" where Mesoamerican community leaders debated issues, formed coalitions, and resolved problems. As poor women sold and served them pulque, the priv-ileged *politicos* in *pulquerías* argued, joked, and haggled until drunkenness gradually shredded the coherency of their discussions. Gallons of fermented agave juices served as their lubricant and, later, their salve. After hours of agave-infused bull sessions, it was difficult to tell whether the *politicos* or the pulque held the most power!

Since the onset of ancient pulque production, agave juices—whether dis-tilled or merely fermented—became fundamental to cultural evolution in Mexico and have continued in that role for well over three thousand years. The Indigenous cultures of Arid America and Mesoamerica were using

agave beverages such as pulque, for ceremonial and sacramental purposes
for at least twenty-five centuries before the Spanish offered them their first
Christian communion. In comparison, altar wine has been used in this
manner in Eurasia for only two millennia since the era of Jesus in Galilee.

⚜

The pre-Columbian significance of drinking agave may be difficult for us
to fathom today, for we do not share the same religious view of alcohol as
was once common in Mesoamerica. Sure, wine is central to the Christian
rite of the Eucharist, but it is not the primary use of wine. When servers
uncork a Bordeaux and pour you a taste, they are unlikely to announce that
it is the blood of Christ. Likewise, when your bartender pours you an ounce
of traditional, small-batch mezcal, they are not likely to begin with chanting
and ablations. The wise *cantinero* likely refrains from declaring what faith
they follow, what person on a nearby stool may be your next lover, or why
you must tear the heart out of your chest if you are to assure a good harvest.

Still, there can be little doubt that for a very long time in Mesoamerica, sacred ceremony was the primary setting for the use of agave spirits.
Like frankincense from Yemen, myrrh from Ethiopia, or musk from the
glands of white-bellied Himalayan deer, a minuscule quantity of the distilled essence of Mesoamerican agaves could vault one into the spirit world.

⚜

That all abruptly changed when the Spanish conquistadores first glimpsed
Aztec ceremonies that they perceived to be fueled by intoxication from
agave-derived alcohol. These imperialistic presumptions and racial ste-
reotyping of the religions of so-called "heathen primitives" led to horrific
oppression, decades of unspeakable violence, and cultural genocide by those
who wished to "take the Indian out" of the man or woman. To no modern
person's surprise, the invaders themselves were plenty brutal even without
a ceremonial sip of mezcal.

The Spaniards responded to what they saw as pagan practices with a set
of rules that suppressed Indigenous rites while supporting a convergence
of the colonists' own economic and religious interests. The conquistadores

banned the agave ferments used in these rituals and reduced the possibility that agave alcohol could compete against Spanish wine imports into the New World.

The Spanish colonial suppression of the initial microdistillation and ceremonial use of agave spirits was both tragic and unfortunate, for it dismissed the sacred and aesthetic value of agave spirits. The Mesoamerican elaboration of both the alcohol itself and its magico-religious uses were cultural achievements as remarkable as the 2,600-year-old *chemer* wine factories in Lebanon, or the scotch whiskey distilleries of Islay Island off the Ireland coast.

These spirited ventures not only are landmarks in human innovation and creativity but express the sacredness and uniqueness of each place in which they were elaborated through their unique terroir. Why should we regard these cultural achievements as any less valuable than the Egyptian pyramids or the Great Wall of China, simply because they were not built of stone, but of succulent plant mandalas raised up into delectable spirits?

Mezcal may be fashionable now, but in the Indigenous villages where it is made, its roots run unfathomably deep. Agaves in Mesoamerica in Arid America are what bison has been, to Plains Indians, or salmon to the fishers of the North Pacific rim: a cultural keystone that serves to meet most human needs—food, fiber, beverage, medicine, construction materials, and spiritual offerings.

In addition to the roasted *penca* leaves and *mezonte* meristems or hearts, nearly every element or organ of an agave has been eaten or taken medicinally in Indigenous communities. Quite wonderfully, these traditions and folk sayings about them survive undiminished today. Mexican cultures have found ingenious culinary means to prepare the very "skin" of the agave leaf called the *mixiote*; the tall flowering stalk called the *quiote*; and the sweet, tender blossoms variously called *bayusas, cacayas, huexotes, golumbos, hualumbos, machetes,* or *xhiveries*. These food and beverage products live on today in at least twenty-four of Mexico's thirty-two states.

If you love *dichos* (folk sayings) as much as we do, listen to what one old Oaxacan proverb recommends to us: *Mezcal: Si te da catarro, un jarro; pero si viene con tos, mejor que sean dos.* "Mezcal: If you have caught a cold, take a cup; if a sore throat comes too, take two."

At his Don Isauro distillery in the Tehuacán Valley of Puebla, Fernando Barragan Flores pointed out to us that different agaves have different medicinal properties: "You know I make *pechuga* using only the *mezontle* hearts of the *pizomel* agave. I value it as cough medicine because its sugars are so dense, almost like molasses. Many mezcals have always been used as medicine, but this one is best for pulmonary congestion and infections. Others work too, especially if they are condensed down, but *pichomel* seems to have the strongest medicinal properties of any we have here."

That reminded me of what Vicente Reyes has remarked about the distinctiveness of each agave in its relationship to humankind: "Each maguey is unique, a one-off, and this is not something that man can or should meddle in."

Months before, as we sat together sampling mezcals in his Lalocura distillery in Santa Catarina Minas, Oaxaca, owner Eduardo "Lalo" Ángeles summed up the agaves' many uses, values, and lessons: "The abundance of gifts that God has given us in the maguey is something that I am deeply grateful for. But this abundance has also taught me how to think broadly about how to properly use these gifts. Agaves are so much more than just a source of alcohol, fuel, or flavor. I've been reflecting lately on their healing powers—their medicinal value—for that is perhaps something that was so pervasive in my upbringing that we took it for granted in my family."

"When I was a child, near every family I knew—including my own—had a little niche in the wall by their door or a little shelf in the kitchen where they kept handy a bottle of mezcal infused with healing herbs. It might be infused with rue, or garlic, or mint. It was readily accessible should someone develop a rash, measles, or a bruise. Of course, we would use it topically, rubbing into sore muscles, or into scrapes and cuts as a disinfectant."

Lalo leaned back in his wooden chair and took a sip of mezcal from a *jicarita* with a beautifully carved design running around its lip.

"We also sipped mezcal as a mouthwash for sore throats, infected gums, a sore tooth, or chapped lips. It can reduce fevers. We grew up never even doubting the healing power of the maguey."

Lalo exchanged with us a litany of other medicinal uses too long to list. In Oaxacan villages not far from Santa Catarina, where water can be periodically contaminated, mothers wiped down the tender skin of babies with mezcal each morning or immersed the baby in a small basin of this disinfecting distillate. We now know that the abundance of sticky, caustic saponins in agave leaves can coagulate blood and slow the growth of tumors. The inulins in roasted agave hearts can lower blood sugars and carry away fats that increase cholesterol levels, while stimulating the steady release of insulin in the pancreas. They have beneficial prebiotic and probiotic effects on gut microbial activity and smooth intestinal function. They can also stimulate blood circulation and reduce inflammation.

After several minutes of exchanging *remedios* of cures attributed to agaves, Lalo seemed to get restless, as if we were not getting to the underlying power that the maguey offers us: "The maguey has taught me things of great value in my life—patience, steadiness, integrity, resilience. . . . You cannot make good mezcal if you do not make a good *palenque* . . . the roasting pits and fermentation tanks, the water and the firewood. . . . To do justice to the gifts of the maguey, you must train your staff well, and work with other people cooperatively. And you cannot support a good *palenque* if you do not have a diverse *milpa*—not just lots of agave varieties, but other crops as well. You need to learn how to best use bagasse that is left over from the roasted agaves, incorporating them into the soil as compost or as mulch to conserve water."

As Lalo spun us further and further out into the agave ecosystem, it felt like his vision of agaves widened to embrace all its relations.

Days later, at Lalo's sister and brother's distillery a few miles away, the Santo Niño del Palenque is perched in a little chapel on a huge boulder that sits above their *olla de barro* distillation patio. The saint is paraded around at celebrations during the Feria de Santo Niño, a festival that occurs in January in many Mexican pueblos. Although Santo Niño del Palenque is not a saint formally recognized by the Vatican, he is a saint in the hearts and minds of Oaxacan *mezcaleros* everywhere.

At fiestas in some pueblos, the saintly child is gifted a small flask of mezcal so that he can be splashed by a mist of spirits before one takes a sip for oneself. He is present when mezcal is drunk to bless a newly married couple or a young woman who has become big with child. And he is there when old *mezcaleros* pass away, so that he can assure they remain in every villager's memory forever.

One only needs to see the omnipresence of mezcal at Día de los Muertos ceremonies all across Mexico and the US Southwest to wonder whether it's the preferred drink of the dead. Little bottles of mezcal are brought into chapels and cathedrals to be offered up to the various *santos* dwelling on the altars. The campesinos tip their bottles of mezcal to their saints before drinking it down themselves, as if *they* are blessing the saints, not vice versa. If you make it past midnight in the streets or cemeteries of Oaxaca, Chiapas, or Michoacán, everything around you seems a bit hallucinatory, as if you've stepped into a mezcal-charged scene in Malcolm Lowry's *Under the Volcano*, and can no longer find the exit door.

Yes, mezcal tastes like time, in that it is both ephemeral and eternal.

In such pueblos, mezcal has insinuated itself into the *dichos* or folk sayings that flavor so many conversations. Innumerable Americans and Europeans who have traveled in Mexico have come home with wooden signs or cotton towels that proclaim the first section of the following *dicho,* while ignoring its equally entertaining second stanza: *Para todo mal, mezcal y para todo bien . . . también. Y si no hay remedio, pues litro y medio.* "For all that is good, [take] mezcal, and for the bad as well, [take] mezcal. And if there is no cure, a liter and a half will work for sure."

The great Mexican ethnographic writer César Augusto Patrón Esquivel has riffed off this *dicho* to explain the deep role that mezcal plays in generating good will in the Mixtec communities of Oaxaca: "The collective participation of the Mixtec in community life begins with the collaborative elaboration of mezcal. . . . It extends through the life of each community member as they share it during religious festivals, seasonal rituals, celebrations, and death anniversaries until the very day that each of them is buried in the ground. Mezcal lubricates and shapes the identity and distinctive culture of each Miztecan pueblo, for it is venerated as a form of alchemy in which all of them can participate, which all of them desire and deserve. When all is said and done, the *mal* (bad spirits) can no longer exist when the presence of the *biénestar* (wellness) comes in the form of good mezcal."

Here's another well-known quip: *El mezcal no te crea ni te destruye, solo te transforma.* "Mezcal cannot make or break you, it only can transform you."

Chapter Five

THE THORNY PROBLEM
OF FARMING AGAVES

Sin *embargo*, agaves are firmly embedded in Mesoamerican and Arid American agricultural traditions, but these traditions are dynamic, not static. In most places, farming for mezcal production has not become so industrialized that it has negated the direct and ancient symbiotic interactions between agrarians and agaves themselves.

If we compare the traditional farming of agaves to any other form of agricultural production used to make distilled spirits, one thing is abundantly clear: The ancient ways of growing maguey have a much lower environmental footprint but a higher level of cultural investment in matching agaves with other plants-—like fruit-bearing tree crops or prickly pear cacti—in intercropped polycultures. They shape a place where perennials dominate.

In such agroforestry systems, human intelligence, skilled labor, and traditional knowledge play a far greater role than any machinery and chemicals in bringing the agave plants to harvest. You can see evidence of these investments in the elegant ways that agaves are integrated with other useful plants in *milpa* fields that have been shaped for centuries by the discerning actions of the harvesters known as *jimadores* or *mezcaleros*.

In the valleys of Jalisco, we have been fortunate to know members of several families that have cultivated agaves for the makers of mezcal at least since the mid-nineteenth century. They draw on local knowledge that often predates by hundreds of years their own generation's roles in the tending and trimming of agaves. Since written documents from as early as 1576 trace the agricultural production of agaves for distilled spirits back to their villages and the surnames still found there, their families' legacies of cultivating plants for the making of mezcal echo through twenty or more generations.

If a single set of faces stand out among all those involved in tequila production, it may be those of the *jimadores*. They are highly skilled agave tenders, cultivators, and harvesters—as important to the taste of tequila as multi-tasking *maestros mezcaleros* are to the taste of other agave spirits. Without a doubt, there are many small-scale distilleries of artisanal and ancestral mezcal where the sower, cultivator, harvester, and distiller are one and the same person.

The *jimador* is as essential to the cultivation, care, and quality of *tequilana azul* or *mezcal espadín* as the cowboy or *vaquero* is to the quality of both delicious beef and productive rangelands in Mexico's ranching landscapes. A *jimador* must be trained well in identifying which agaves are fully ripe and ready to harvest, for a prematurely harvested agave may render a bittersweet or sour taste, ruining an entire batch of distilled spirits. If the neglect of weeds or diseases weakens a maguey, the resulting spirit might not be as robust in its sweet inulin content and heady aromas. The plant can take the heat, but infestation by weeds or root diseases weakens its capacity to absorb minerals and express its volatile oils.

Of course, there are always several other livelihoods involved in assuring that a distillery is making good mezcal, but the care for the materia prima matters. So, the role of the harvester is fundamental, like the person steering the rudder of a raft. At their best, a *jimador* who tends, trims, and harvests the many kinds of agaves found in agrarian landscapes can add value to a mezcal.

Over decades if not centuries, the iconic position of head *jimador* for a work crew doing the *jima* (harvest) and *barbeo* (pruning or trimming) of agaves

The jima *pruning of agaves with* coa *by a* jimador

has been passed on from one generation to the next through the same extended family. The crew leader trains all the *jimadores* in how to apply at least five techniques for trimming and cleaning agaves at different stages of the life cycle.

The first technique is the *barbeo de arborolito* or *farol,* to trim leaves to promote better aeration and prevent diseases, pests, and stress from competition with weeds. The second is the *barbeo de cacheteo* or *chaponeo,* to gain access between rows of spiny leaves. The third is the *barbeo de desemplague,* to reduce the impact of the *gusano barrenador,* a larva that damages the leaves. The fourth is the *barbeo de escobeta, rebajado,* or *castigado,* to trim all leaves to increase solar gain and accelerate maturation. The fifth (before harvest) is the *capona or desquiote,* to "castrate" the *quiote* flower stalk as it readies to blossom.

Listening to a longer description of the subtle differences among these pruning techniques might disorient you at first, but we are willing to bet you can taste the difference between the three harvest techniques: *jima razurada, jima normal,* and *jima larga.* In *jima razurada,* the waxy, spiky leaves are removed as near to the *piña* as possible; it is clean shaven. In *jima larga,* a considerable amount of leaf matter is left on. The techniques are chosen by *jimadores* based on their perceived effects on the sugar content of an agave: The leaves are often bitter, so the core of an agave bursting with sugars can be balanced with natural bitterness using the *jima larga* technique. Conversely, an agave low in sugar would be out of balance with the inclusion of longer leaves, and *jima razurada* would likely be chosen. Seasoned *jimadores* assess what trimming technique will be best for sugars and flavors of the batch.

The unskilled *jornalero*—a daily wage worker—would have a hard time learning how to finesse these trimming techniques from an experienced *jimador* overnight. They must master three ways of achieving final *barbeo* for different objectives. They must make a perfectly shaped pineapple-like *piña* out of a *cabeza* or *caballo* of the mature agave. How well they do so will have significant consequences for the harvest weight as well as the quality and roasting of the agave. They do not learn these techniques with their heads alone; they master this tactile art by feel and through internship and deep engagement, not by abstraction.

Historically an inherited position, the status as a *jimador* crew leader is often bestowed upon a particular family member who is both muscular and extremely precise in trimming and uprooting agaves. While most of those chosen for that position have been men, make no mistake that women have been honored as well.

To be sure, some very good women have also mastered the art and science of the *jima,* although their presence among the largely male *jimadores* is often overlooked. In addition to being superbly skilled with using the tools of the trade—various *coas,* machetes, and hoes—the *jimador* must also be socially adept at keeping a crew engaged as it works long hours in the sun amidst the thorny, spiny, sword-shaped leaves of maguey, with *piñas* known to cause persistent itch when exposed to bare skin. To put it bluntly, no lead *jimador* wants to miff a crew of guys with razor-sharp tools in their hands. These elegantly hand-forged tools are lovely to look at, but deadly to mess with.

From dawn until three in the afternoon, the head *jimador* must know when to command, cajole, or humor the crew. He must discern whether to offer more training, to give first aid, or to fire a reckless worker. He also must know when a weak or damaged plant must be removed and how to replace it with another. He must remind the members of their crew that the way they care for the earth at their feet can improve soil fertility and flavor in the mezcal or leave more plants weak and vulnerable to disease.

Above all, the *jimadores* must impart such traditional knowledge about managing mezcal not only to the crew, but to the distillers' consultants, who may wish to cut corners on agave care to reduce the costs of production. If anyone must ethically hold the line on maintaining the tradition, it must begin with the *jimador*. If mezcal production is guided by an unwritten library of traditional agronomic knowledge held among the harvesting crew, then the head *jimador* is the librarian, the archivist, and the keeper of the oral histories. Recently, *jimador* Armando Acevez alerted journalist Hannah-Ellis Petersen of *The Guardian* that this unbroken chain of intergenerational inheritance of *jimador* know-how is sadly breaking down: "Almost every man in my family has been a *jimador*, going back generations, it is a tradition. As long as I can harvest I will keep being a *jimador*. But most of the *jimadores* now are old people because none of the young people want to work on the harvest. You find a few young *jimadores* but there are less and less."

Sadly, as a Jaliscan agronomist named Rubén confided in us, with the loss of this traditional knowledge before or during the tequila boom, "The management of *tequilana azul* plants in fields had a profound effect on the quality of the materia prima." There were more fields and more hired hands, but fewer seasoned *jimadores* to assure quality control.

You can sense this difference in the stance, dress, and conversations of the older workers who grew up in the industry versus the younger hired hands from outside the region. In the old days, the traditional *traje* or garb of the *jimador* included white cotton shirts and pants with red sashes, *tapatio*-style sombreros, leather gloves, and open-toed leather sandals. Today, the *jima-*

dores and younger *jornaleros* dress much as working cowboys do, with heavy boots, blue jeans sometimes covered with protective "chaps," tight-fitting gloves, and cowboy hats of straw treated with a weatherproof sealant. They may wear headphones or earbuds while they work, listening to the latest *narco-corridos* rather than the old mariachi tunes of Tapatian charros.

Still, it is no exaggeration to claim that the skin of *jimadore*s is often as leathery and torn as their boots and gloves. Their faces are wrinkled and bronzed by the scorching rays of the sun. Their hands, calves, and thighs may show old scars from accidental cuts from the blades of the *coas de jima* they use for trimming and harvesting the pineapple-shaped hearts of agaves.

Jimadores are often portrayed in a romantic, almost nostalgic manner as handsome but rough-hewn faces of the agave industry, like the iconic Marlboro Man who knows outdoor environs as well as the back of his hand. A few elderly masters of this art are put "on show" whenever the tourists come to a fancy plantation to see where and how their favorite drink is made.

But as the tourists leave the scene, the *jimador* may limp away, holding his *coa* like a steadying cane. He may be put on show because he can no longer harvest two hundred to three hundred fifty agaves over the course of a day, as he did when he was a kid. He still has the eye to precisely cut the leaves off the agave in one of three forms of *barbeo*, but he may no longer be able to heave 70- to 100-pound (30- to 45-kilogram) *piñas* of agave onto a flatbed truck for hours on end. And it seems that the pay he gains cannot keep up with the pain he must endure, so he wavers on whether he should throw in the towel. Like the making of mezcal itself, it seems that some level of suffering and pain goes into the making of a top-notch *jimador*, who wears his scars with pride, like a badge of courage.

For all the romantic notions and nominal gestures of respect heaped upon the *jimador* as a distinguished profession passed on from father to child, they are never given the full status of a farmer, because they typically do not own the land where the harvests of agaves occur. Many of the farmworkers are those who joined the ranks of the dispossessed, early in their twenties.

✼

Commonly, small landholders now rent out their lands, for they have little interest in being among the fourth or fifth generation to be wizened or wounded by a half century of hard labor. They buy cars or renovate their homes with the rental fees, but now live in Guadalajara or Tequila, working as tourist guides, taxi drivers, herbicide salesmen, or barbers. They have given up cutting spiny agaves to pursue less strenuous occupations.

Today, the land is increasingly owned or rented by the large multinational corporations that run the biggest distilleries and distribution networks. Yes, there remain some independent growers with 15 to 20 hectares whose family members are among those who labor in the agave plantations, but their numbers are dwindling. Few wish to participate in the *jima* itself, even on their own land.

✼

Curiously, the Mexican term *jimador* does not exist in other Spanish-speaking nations. Historically spelled *ximador,* this native Mesoamerican word usually refers to a master harvester, cultivator, pruner, or even "plant barber" rather than to a farmer or *potrero* owner per se. It is an adaptation of an ancient Nahuatl term of the Aztecs, *xīma,* "to smooth into a desired shape, trim hair, shave a beard, prune a shrub, plane a piece of wood, or sculpt a piece of stone."

The *jimador* is the critically important taskmaster, shaping the wild mandalas of agave leaves into a compact and often stackable set of *piñas,* trimmed to look like unblemished and disease-free pineapples ready for roasting. No healthy *jimadores*, no healthy, robust agave heads. It's that simple.

Once in Atotonílco, a tequila-producing town in the Los Altos region, David spotted a highly regarded *jimador* who was struggling to get down off the bed of a pickup truck.

"Are you okay?" he asked the *jimador*. "Did you just have an accident?"

"No, it's chronic pain from so many injuries. And I'm out of cash to get more pills. When I'm not taking pain killers and muscle relaxers, I can hardly work anymore."

The careful cultivation of agaves—as opposed to merely encountering wild ones in natural habitats within mountain ranges or rolling plains—requires that consumers and society at large care for the health of those best able to do the arduous work of a *jimador* or *jornalero*. We need their steady stewardship and collaborative interactions with other workers if we are to partake in their quality spirits without any regret. Otherwise, agriculture will further devolve into an extractive industry that depletes not only the soil and plant life it depends upon, but the workers themselves. Even in the most mechanized monocultures, they remain essential to every agricultural task, from sowing the seed plants to reaping the harvest.

More than ever before, the issues of sustainable farming and worker justice are hitched at the hip. Fortunately, they are being widely discussed within the mezcal industry, in cantinas and in journals. In some quarters, it remains risky to raise such issues, as it is the proverbial *pedo* (fart) in church! But both the workforce and the sustainability of Mexico's food system are facing dire dilemmas. Now is the time to solve them.

As an embodiment of the mutualistic relationship between man and the maguey and the long, slow dance through time and space that defines Mexico and its history, the harvester is not merely the harvester who goes out on hillside terraces with his mule for a quick machete swing at an agave. He is also the planter of the next generation of agaves that will fill the space left by the trimmed and uprooted *piña*. In turn, that exquisitely trimmed pineapple-shaped agave offers him his wages and, later, nourishment, inebriation, and pleasure.

As a mature agave begins to send up its *quiote* to burst into hundreds of pale blossoms, the harvester must decide whether to *caponar* (castrate) the stalk to let the plant swell with carbohydrates before it is harvested for mezcal. As another option, he may allow the flowering stalk to fully develop into a candelabra 10 to 30 feet (3 to 10 meters) tall. This is done in order to let the plant season and accumulate more flavor, fragrance, and sweetness as the

Tahona *grindstone crushing agaves in Egyptian mill*

roots take up nutrients from deeper soil profiles and more aromatic secondary chemicals become embedded in the leaves and heart.

Some *maestros mezcaleros* like neighbors Don Macario Partida and Don Lorenzo Vírgen have the restraint to let a stalk and the heart below it season for up to four years, a patience you will not soon witness anywhere in tequila country. Alternatively, the harvester may decide to cut down the stalk only after the bat-pollinated flowers produce seeds. The viable seeds will then be germinated in a nursery so that the seedlings can be planted months later in the terraced *potrero* where they will mature.

After a particular agave has been growing for eight to thirty years in the harvester's presence, a kinship has developed among the two. The harvester may go out to the seasoned elder in his fields, and tap the flowering stalk to see if is drying and hollowing out. He may look for the blush in the agave's leaves to see if the sweet fructans have been mobilized. Each agave has its own personality, its own pace to reach the stature of *sazón*, seasoned maturity. As the wise agronomist and distiller Carlos Camarena of the Tapatio distillery once quipped, "We live in an age where we want to rush through everything, but why not take some time to sit with one another, to *convivir?*

That is something the *jimadores* and *mezcaleros* routinely do in Oaxaca, but when we are in the United States, we just rush around looking for all the answers. No time for one-on-ones!"

That is exactly why the blue deserts of tequila monoculture in the valleys and highlands of five Mexican states seem more like an anomaly today in the agave world than a venerable tradition. They signal a rush to slaughter. Perfectly manicured, even-aged stands of a single clone seem more like a bad parody of a barbershop than the way to honor the ancient Mesoamerican symbiosis.

There is something about tequila monocultures that makes us blue in the face. It is as if we are holding our breaths for a sign of relief. The very notion of fifteen to twenty million genetically identical individuals of any species planted out every year across a landscape of nearly straight rows over 300,000 acres of land sounds more like a scene from a dystopian science fiction novel than a dream for a healthy and rewarding enterprise.

But don't worry too much—help is on the way.

Chapter Six

✹

THE MEZCAL MASH

Together, we watched as the *tahona*—the enormous millstone—rolled around and around in what is called an "Egyptian mill." It was crushing dozens of roasted heads of agave at the Cascahuín distillery. That is where agave spirits have been made by Salvador Rosales Briseño and his descendants in the rural town of El Arenal since 1955. And that's where the Rosales family has installed a *tahona*, becoming one of just a handful of distilleries that still use the 2-ton stone to crush and enhance the flavors of the agaves oven-roasted on site. The mating of stone and plant releases sticky juices and sweet, smoky pulp; it is hard not to be entranced by this ancient circular dance between geological forces and botanical biomass as the stone renders hints of its mineral, earthy fragrances. And when sugars are liberated from the *penca* fibers during the crush, organic acids emerge that add to the sweetness and floral fragrance of agave spirits.

In the "blue desert" lowland of Jalisco, this traditional method for crushing the heads of baked agaves for tequila and mezcal had all but disappeared before a minor revival began just a few years ago. We witnessed an old *tahona* put back to work in El Arenal, but now it is one of many being reemployed in Jalisco.

As the millstone rolls around in a stone-lined basin, hitched to an axle-like rod that connects it to the turnstile in the middle of the basin, it presses the aguamiel or golden juices out of the baked leaf bases of the *piñas* being crushed. Along with those juices, gutters edging the mill basin capture the pulverized pulp of the *piñas* and funnel them away to fermenting vats.

Their dreamy aromas linger on, pervading the air in the courtyard of the distillery. We had both grown up seeing such contraptions work in our different parts of the Americas. David saw them while still an adolescent in western Mexico, where they crushed agaves and many other plants for syrup. Gary had seen similar apparatus as a child in the US Midwest, where farmers pressed juice out of sorghum stalks to make blackstrap molasses; and in his twenties, to extract the sweetness from both agaves and sugarcane in Sonora. A tahona emits an ancient sound as it rolls around. . .

The term *tahona* harkens back to Andalusian Arabic *ṭaḥúna*, derived from classic Arabic *aṭṭāḥūn[ah]*, and it simply means "mill." Gristmill, millstone, all the above. Its origin, interestingly, has nothing to do with the Aztecs, who likely used wooden mallets for the crush of agaves and other plant foods. The antiquity of *tahonas* in both the Mediterranean and Middle East goes back many millennia, as Gary has witnessed in his ancestral village near the Lebanon-Syria-Israel border.

In fact, the *tahona* was likely introduced into Mexico for use with multiple crops in the sixteenth century, perhaps by crypto-Muslim and crypto-Jewish refugee families escaping from the Spanish Inquisition. Andalusian millers called *tahhán*—another term derived from Arabic—still pass on the trade from generation to generation, grinding grain, mashing grapes, and pressing sugarcane with millstones turned by draft animals, just as many rural Mexican families have continued to do well into the twentieth century.

The only Indigenous element to this mestizo technology is that the preferred rock type for making *tahona* millstones in the Trans-Volcanic Belt of Mexico is called *tezontli*. That Spanish term is a joining of two Nahuatl words, *tetl* for stone, and *tzontli* for hair follicles on the head. It describes the highly oxidized but rather porous amygdaloid stones easily quarried from volcanic strata in west-central Mexico. The same quarries were probably used to obtain reddish or blackish rock for making spice-crushing *molcajete* grindstones in pre-Colonial eras. We'd have no *moles* without them.

※

With modernization and the mezcal boom, many *mezcaleros* and *tequileros* have abandoned both their volcanic millstones and their wooden mallets for speed- and volume-focused mechanical roller mills and chipper-shredders. While the use of *tahonas* in the crush of agaves persists among a small percentage of *mezcaleros* working from Sonora clear to Oaxaca, only a handful of tequila distilleries kept them active during the late twentieth century. Then Carlos Camarena of El Tesoro Tequila dusted off a well-cured 3-ton grindstone he had inherited from his grandfather, who in turn had acquired it from his own grandfather. As Carlos describes: "We are prepared to do this work with our hands and hearts. This is what the *tahona* means to us. It means our hearts."

Thanks to skilled traditionalists like Carlos, Salvador Sr., and his son, Salvador Jr., the use of the *tahona*—pulled by mule or tractor—has been revived for the making of ancestral mezcals and tequilas. But what might it add to an agave distillate? Is there a discernible enhancement of flavor and aroma?

Our friend and agave spirits specialist Clayton Szczech sees the use of the *tahona* as a symbolic commitment to tradition: "The *tahona* is more of a signifier of a commitment to a quality process, regardless of slow speed . . . rather than a technique that does anything essential."

Others, including David, respectfully disagree. Portland's whiskey wizard Tommy Klus senses that the stone-ground crush really does add discernible flavor: "It takes me back to the distillery, as I recall standing at the ovens as they were about to unload the freshly roasted agave." Mariana Sánchez Benítez, who oversees production at a Patrón distillery senses that "*tahona* tequila tends to have a smoother and sweeter profile." She conceded to Jake Emen, "It is a very, very slow process. Efficiency suffers, but we can retain the flavors we want."

Depending on the species of agave—as well as the roasting time for the mezcal—the flavors enhanced by *tahona* processing seem earthier, with more minerality. Some *maestras* claim that they are more vegetal and herbal as well, for the crush bruises rather than obliterates the many layers of any agave leaf. This may make the taste of a *tahona*-crushed mezcal richer and rounder on the tongue, with a more sweeping presence of the distinctiveness of the particularities of each agave showcased in a longer light.

But is that why many *mezcaleros* are sticking to *tahonas* or mallets for the crush? It seems they are also doing it as an act of resistance against the most consequential—and controversial—innovation in the history of tequila production: the diffuser.

Before the diffuser, most mezcals and tequilas were processed this way: The harvester selected mature agaves to be harvested when they were swelling with sweet fructans, which he shaped into *piñas*, sometimes halved, and then roasted. The roasted *piñas* were then chopped up by mallet or millstone, then fermented, distilled twice, and bottled. Historically, this entire process took a minimum of a week to complete, but typically many days more to get the complex taste profiles that *mezcaleros* adore.

Then the diffuser came along. A diffuser is an industrial machine the size of a regulation basketball court that is essentially designed to turn whole agave *piñas* into goop. It flips traditional production on its head by shredding the raw, often immature agaves, blasting them with acid and scalding hot water from high-pressure jets, and cooking the remaining slurry in an autoclave. The slurry of sugary juices is then fermented with chemical accelerators, and distilled—not as tradition intended.

The benefits of this short-cut method are that producers can churn out massive quantities in a manner of hours, down from a week or more in traditional production. Because Mexican law now allows for the addition of extra, non-agave sugars, colorants, and artificial flavorings, there is no real need for those committed to this production style to wait for agaves to mature and develop their natural sugars or flavors.

Put aside the fact that great organoleptic complexity comes with maturity and is the hallmark of distillates traditionally composed of the venerable cen-

Baked agaves cooling after tatemada *in pit*

tury plant. Instead, whatever flavor the immature agave may have accrued
is all but lost in "flash processing." Even if a perfectly ripened roasted agave
were put through a diffuser, its end product would taste as vacuous as that
derived from an immature agave. The industrialized end products of both
are more like the high-fructose corn syrups found in thousands of processed
foods than like the less-fragmented fructans and inulins from mature, slow-
roasted agaves. And technically, these end products should not be called
mezcals at all, because they are not truly *mex-calli*—roasted agaves.

Nevertheless, we are told by the industry not to worry! The pleasure principles—aromas, flavors, colors—can always be artificially added by industrial
food chemists later. Well. Maybe. Between the diffuser and the autoclave,
agave matter is bludgeoned by unbelievably rapid and nutritionally destruc-
tive physical and chemical processes. As a result, any finely crushed plant
fragments from young *tequilana azul* or *espadín* agaves are pulverized in
a power wash of hot water stoked with acids and enzymes. This forcefully
removes all lingering flavors worthy of any note that were once naturally
embedded in agave's marvelous inulins. In short, fast, cheap, and easy is
rapidly replacing slow, rich, and careful at the scene of the crush.

Chapter Seven

※

BLESSING THE FERMENT OF AGAVE-LOVING MICROBES

One beautiful spring day at the 2022 Agave Heritage Festival, we asked Mexico's greatest expert on the evolutionary ecology of microbes, Dr. Valeria Souza Saldívar, to help our fifty fermentation workshop participants place agave fermentation into a larger perspective. As we sat together on the patio of Amy and Doug Smith's El Crisol mezcal bar, Valeria wasted no time in weaving a parable that none present that day will ever forget: "In the beginning, the Earth was much like a ball of clay sailing through the universe, orbiting our sun. But rather suddenly, a comet collided with the Earth, and turned it into something more akin to an *olla de barro*—a clay pottery jar. The moisture on Earth drained down into the jar, brewing up the Comet Soup that gave rise to all life on this planet. That's right, life began with a process with similarities to microbial fermentation. We would not even be here were it not for those microbes. Our ancestors emerged from the blessed ferment! It may have happened on a very large scale, but it is not all that different than what happens in a fermentation vat filled with roasted agave, moisture, and microbes! This is something to celebrate: our debt to and kinship with microbes!"

That may be the sweetest and headiest creation story we have heard, but it is not merely one to imagine with your eyes. Just try to imagine it with your nose and mouth as well!

When you first walk toward the underground, stone-lined fermentation tanks where mezcal is being produced, you can smell their aromas before you see them. At Maestro Santos Juárez's *taberna* in Canoas, Jalisco, the fragrances emanating from the fermenting heads of baked agave are earthy, savory, potent and provocative.

As you arrive at the edge of the stone-lined pits to look down into their murky waters, what you see looks like a *mole negro* sauce with a lot of shredded turkey or chicken floating in it, but from the aromas that reach your nose and mouth, you might as well be sniffing the primordial soup where life first evolved. You can pick up some sour notes, but there are also the familiar fragrances of citrus fruits, lily of the valley blossoms, cocoa, almonds, mints, malted grains, and pine wood.

The strange brews that bubble up, froth, and fizz from the underground vats of mezcal-making *tabernas, palenques,* and *vinatas* across Mexico are extraordinarily rich in microbes that stay at work for as much as a month converting the carbohydrates of agaves into the beverages we love to savor.

In two small regions of mezcal production in the state of Durango, microbiologists have isolated more than 170 strains of yeasts and dozens of bacterial strains from roasted, fermented agave musts. These unseen and sometimes still-unnamed microorganisms convert about 80 pounds of freshly roasted agave heads or *piñas* into what will become 10 gallons of distilled mezcal. They are the alchemists who call up the spirits in the maguey.

For most mezcals, the first step toward fermentation occurs in stone-lined roasting pits called *hornos*, where the red-hot embers of charcoal from selected firewood and volcanic stones retain heat in the pit for three to ten days. The layer of stones and tarps covering the pit is then removed, and the golden-brown heads that have survived the *tatemada* or slow pit-roasting are allowed to cool in the open air for twelve to thirty-six hours. It typically takes anywhere from 20 to 45 pounds (10 to 20 kilos) of trimmed wild agaves to render a liter of 100 percent agave distillate, but the range varies considerably among the species used. Once the sugary liquids and pulp are extracted, more than 33 pounds (15 kilos) of fibrous bagasse remains to be dealt with as a by-product. Unfortunately, it is too often discarded, rather than being used as mulch in fields, composted, or fermented for use as a livestock feed as the best producers do.

Next comes the crush that we spoke of briefly in the last chapter. It can be a laborious task. The roasted heads are either crushed by the stone wheel pulled around by a beast of burden in a circular basin called a *tahona*, mashed with a wooden mallet called a *mazo*, or run through either a steel-tined roller mill or a mechanical chipper-shredder called a *trituradora*. In many regions, the *tahonas* remain key to making ancestral mezcals, as currently defined.

After the crush is over, the lesser skilled, younger members of the crew sort the roasted fragments into three groups: the golden, finely baked; the deeply brown and caramelized; and the black, badly burnt *pencas, quiotes,* and *corazones.* The burnt pieces are kept out of most fermentation vats, but may be shredded again and then mixed with clay to cover the clay pots used for ancestral distillation.

The larger, more salvable pieces might be cut, crushed, or mashed by *mazos* or axes in a shallow stone basin called a *canoa* or *tina*. This is done to further reduce their particle sizes and maximize their surface-to-volume ratio for the next step of processing in fermentation vats. Today, more and more chipper-shredders are used for one step of the crush to reduce labor costs.

It is the third step that takes the most time, but it is most critical to the flavor and fragrance of the product, even more than the care put into distillation itself. It is the slow fermentation you saw when you reached the underground vats—the mixing of specially selected water with the roasted, mashed fragments of the agave heads. Vats may be made of any kind of nonporous material, but commercial *mezcaleros* often make barrel-shaped tanks that hold volumes of 800 to 1,200 liters. Much of that is water, but not just tap water: Spring water with high minerality and earthy sweetness. Rainwater caught on roofs or in cisterns. The trickling thread of crystal clear water draining out of sandy seeps. A mountain stream draining down from a forest of the sacred fir known as *oyamel* and the sagging pine called *pino triste*. All leave their marks in the making of agave spirits.

But all of these waters are not regarded as equal after they enter the mouths of *mezcaleros*. Some sources they treat as holy water suited only for mezcal used in Indigenous ceremonies, as if it has rained down like sweat from angels or saints. Others are put to work every day in agave distilleries, as though they are safe but steady commodities.

At this stage, the blessed ferment begins the miraculous metamorphosis of agave matter into spirit. It takes at least thirty hours for the fermentation microbes to work their magic to transform all the carbohydrates into the seventeen forms of alcohol that will later be distilled from the must. And yet, most *mezcaleros* keep their fermentation magic in the vats for far longer than thirty hours—for some, up to thirty days. It usually takes many more hours or days of alchemical emergence from the fermentation vat to let the countless volatile compounds in the agave be fully expressed as flavors and fragrances.

Stone-lined underground vats are not the only vessels used for agave fermentation. Clay pots, wooden *barriles* or *canoas*, carefully sown rawhide vessels, plastic *tinaco* buckets, and stainless steel vats or tanks are also used. The hides from cows and bulls are still used in Guerrero, Michoacán, and Puebla, but are especially favored by Indigenous Oaxacan *maestros* in the Mixteca Alta. *Maestros* from each region have their own vocabulary for pulling up fruity flavors from the lactic acid bacteria that naturally occur in their bags and columns of ferment.

Some *maestras* and *maestros* are extremely picky about the kind of wood they select for their best *bateas, barriles,* or *canoas.* They might prefer bald cypress, willow, oyamel, or aged white oak, since the tannins and volatile terpenes in the wood itself can either enhance or skew the flavor profile. In the French white oak barrels preferred for aging of many tequilas and mezcals, more than sixty volatile compounds enhance the flavors, colors, and aromas of *añejos* and *extra añejos*.

✳

The miraculous metamorphosis of the dark, chunky, musky Comet Soup into a clear distillate happens by way of a multistep process that scientists have only recently come to fully understand. First, the harvested heads of mature agaves or *piñas* must be roasted to hydrolyze their fructans into simpler sugars. Fructans are the principal carbohydrates found in all agaves, no matter what species. They are the "right stuff" from which mezcals are made.

Although you may not have heard of inulins much before, you have certainly consumed them in other plants—garlic, leeks, jicama, Jerusalem artichokes, and chicory roots. Inulins belong to a larger group of polysaccharides called fructans that are probiotics rich in soluble fiber. After starches, fructans are the second most abundant storage carbohydrate in nature. They are fortunately found in many drought-hardy plants—not just agaves—that cover deserts and other hot, dry landscapes. The kind of inulins in agaves are so special that they were recently given their own name: agavulins.

Two features make agave inulins unusual. In nature, they slow down water loss by tightly holding on to the energy and moisture conserved in the storage tissues of the roots, stalks, and "heart" of agaves. But in the human metabolism, these same inulins slow down the digestion and absorption of plant sugars in the gut. They do so in ways that control blood sugar levels in diabetics, while improving digestion, promoting weight loss, and relieving constipation as well as fat buildup in the liver. These qualities give roasted agave heads and their slow-cooked juices the distinction of being "slow-release" foods now sought out by disciples of the Slow Carbs Diet or South Beach Diet.

✳

While some of us once thought that the "magic" was mostly in the genetics of the agave used, or in the skill of *mezcalero* himself, we now realize that the best *mezcaleros* intuit that their job is also to recruit a whole suite of flavor-enhancing allies to help enhance the taste of their spirits. Those allies are the essential workers in mezcal fermentation, for they comprise an extraordinarily diverse set of microbes with the power to heighten the intensity of flavor and fragrance in mezcals or, if improperly managed, to mute it. Like most good things in life, it's all about building and sustaining relationships.

You'll recall that the spirits of mezcal or maguey are definitely *not* distilled from "cactus juice," nor are they made from the aguamiel sap that wells up in the hollows of a castrated giant agave managed for pulque production. Agaves are not even close kin to cacti, although both are sometimes called succulents and share the water-conserving metabolism that we will describe later. Mezcals are brought to you by the rather wondrous symbioses that have emerged from agaves and the diverse microbiome embedded in their roots, leaves, and sap.

Think of it this way: Mezcal is a spirit distilled from roasting and fermenting the trimmed, pineapple-like heads of any of several dozen kinds of agaves that have grown in symbiosis with hundreds of mycorrhizal fungi, bacteria, protozoans, and nematodes associated with soils. These underground alliances give each agave symbiosis its distinctive profile of terroir. And the terroir of no distilled beverage in the world is more a product of life's symbiotic relationships than that of mezcal. It's our Comet Soup of the day, as Valeria Souza Saldívar might say. But each microbe can be a key that unlocks another flavor in that soup.

⁂

Let us follow that intriguing notion down to the specifics of terroir with distillates from Mexico's Pacific coast. Sporadic volcanic eruptions from the Volcan de Colima may spurt out lava and ash for thousands of years, enhancing the soils on its flanks with potassium, iron, magnesium, and phosphorus. These soil nutrients not only fuel the growth of agaves but are also credited with offering volcanic flavor profiles to the mezcals. The ash favors certain kinds of soil microbes over others, which differentially absorb certain minerals and flavor precursors. A nearby sea brings salinity inland in fog and mist that condenses onto agave leaves and drains down into soil. That is what offers such a sharp, briny note to a coastal *raicilla*, just as it does to the oysters in the ocean. Coastal *raicillas* also carry with them more tropical, citric, and smoky notes than do the more lactic *sierra raicillas* from the interior.

But what about the terroir brought in by those who tend the agaves and the soil?

What about the men and women of the pre-Olmec *Capacha* culture who may have begun with agave agriculture, roasting pits or kilns, and ceramic vessels for alcohol production during the Formative period in western Mesoamerica?

What about the agave gatherers and hunters before them, who may have begun the cultural selection of agaves that later led to their widespread cultivation?

How did they influence the growth rates, the sizes, scents, colors, and flavors of the agaves they favored for mezcal?

What about the ways the agaves and their symbionts influenced the coevolution of humans and the microbes in our guts?

After millennia of subtly nurturing the terroir of their plant foods and beverages, how can we separate agave stewards from the soil, from the microbes, or the mezcal?

Why should we ever again ignore the intelligent decisions, the precise labor, the generational wisdom, and the dedication our predecessors—both human and other-than-human—offered to this symbiosis?

If any one of us without that wisdom was to take over the land, the plants, and the stills today, could we instantly produce a distillate of commensurate quality?

No way. We could read all the books about cultivation, fermentation, and distillation, but the resulting spirits would not be anywhere near as extraordinary as the ones we would be regrettably replacing. Even if we let the best agaves be distilled by the most successful spirits enterprises in the world, it is unlikely that they could suddenly emulate all the moves a traditional *maestro mezcalero* makes in their intimate dance with the agave.

The reason for that is simple: These people have been doing this for thousands of years, guided by traditional knowledge passed on through daily practice under the guidance of mentors, not through printed recipes in a cookbook or formulas in a chemical engineering manual.

Nobody else really knows how to do what the *maestras* and *maestros* have learned by practice, prayer, and persistence. They do not perform their magic merely for the purpose of profit; they do it out of pride and culture. What is more, these cultural contributions to the terroir of a mezcal cannot be recreated at any other scale.

※

Now, if the term *symbiosis* seems way too nerdy and clunky to circumscribe all of this for you when you are drinking mezcals and conversing with your friends, blame our late great friend Lynn Margulis for coining an even nerdier, more obscure term. In 1991, Lynn coined the term the *holobiont,* shorthand for a cohesive assemblage of a plant host and the many other species living in or around it, which together form a discrete ecological or gastronomic universe unto itself. Having been married to astronomer Carl Sagan for years, Lynn at first likened holobionts to stellar constellations and to cyborgs. But now, as Valeria Souza Saldívar reminds us, we need to look no further than our own skin and stomach to realize that we too are holobionts. We would do well to more frequently turn to the pronouns *we* and *ours* for whatever endeavor we engage in.

※

These holy alliances between multiple biological kingdoms are what gives agave spirits their distinctive verve and herbality, for they offer a complexity of flavors, fragrances, and textures unlike any other distillate ever imbibed by humankind. In fact, the brilliant Oaxacan beverage chemist Araceli Vega Guzmán has conceded that "the composition of the aromas of ancestral mezcals are extremely complex," undoubtedly more complex than the flavor and fragrance compounds found in any other distilled beverage studied to date.

※

No other spirit in the world comes from so many plant species intertwined with so many microbes as mezcal. For fermented and distilled beverages of all kinds, the proof is in the yeast. Traditionally, most beer brewers use one of two kinds of beer yeasts to provide distinctive body and flavor: the "top-fermenting" ale yeast known as *Saccharomyces cerevisiae* or the "bottom-fermenting" Carlsberg lager yeast now known as *Saccharomyces uvarum*.

Until recently, that has been it for most modern beers. Two species from the same genus, albeit many strains. We can grant that vintners are not as exclusionary as *bier meisters*, for they mostly draw upon *S. cerevisiae* or *S. bayanus* but often allow other wild yeasts to enter the fermentation vat as well. The wild upstarts of the wine industry have bubbly names like *Kloeckera, Saccharomycodes, Schizosaccharomyces, Hansenula, Candida, Pichia,* and *Torulopsis*.

Still, only six kinds or "genera" of yeasts insinuate their way into all the world's wine cellars. For most whiskeys, scotches, and ryes, the yeasts are mostly strains of the same *S. cerevisiae*, though a few secondary species have always been used to kick up the flavor in artisanal whiskey distillation. For rum, industrial fermentation typically relies on just two groups: the various strains of *S. cerevisiae*, together with a few species of *Schizosaccharomyces*.

The fermented slurry for most distillates (including many tequilas) are now brewed up in hyper-sanitary stainless steel tanks using only highly uniform, bioengineered, patented strains of yeasts selected and purified in antiseptic laboratories. In contrast, the musts in the underground vats of stone, rawhide, wood, or lime-stuccoed basins used for mezcal fermentation are as diverse and unpredictable as the spirits they eventually yield.

Thousands of strains of more than thirty yeast species and sixty-five species of bacteria are found in the fermentation vats for traditional mezcals and pulques in central Mexico alone. If we could systematically sample all the fermentation vats in all the *fabricas, palenques, tabernas,* and *vinatas* in every reach of the republic, we would certainly encounter as many bacteria and yeasts in agave beverages as those found in cheeses, kimchis, and kombuchas all around the world.

As such, mezcal can be defined as the only set of distilled spirits in the world derived from a biologically diverse holobiont formed from different agaves, yeasts, and bacteria in true symbiosis.

※

Some of those agave-loving microbes coevolved with their hosts to be endophytic—with their cells living entirely enmeshed within those of their succulent hosts without causing any apparent disease or death-causing parasitism.

One such agave endophyte is a lactic acid bacterium with the euphonious name of *Leuconostoc mesenteroides*. It is found in fermentation tanks for both pulque and mezcal fermentation, giving the former its characteristic flavor and viscous texture by the way its enzymes interact with agave inulins. That "pulque-imbued" flavor profile carries over in the distilled aguamiel of *comiteco*, a beverage unique to Chiapas. Although the distilled *mezcals* and *comitecos* lose the slimy viscosity of pulques, the characteristic taste of *Leuconostoc* still echoes through these spirits.

In contrast to *Leuconostoc*'s pivotal role in pulques and *comitecos*, six yeasts other than the common beer strain dominate the fermentation process of most mezcals: one species of *Kluyveromyces*, another of *Pichia*, a third of *Torulaspora*, a *Clavispora*, and two species of *Zygosaccharomyces*. But in Oaxaca alone, another twenty-eight species of yeasts bubble up at various moments in the fermentation process, each of them adding to the sensory complexity of some batches of mezcals but not others.

Each strain and species of yeast in the fermentation vat offers some value that plain old beer yeasts from *S. cerevisiae* cannot even dream of. In fact, some microbiologists devoted to mezcal chemistry have proposed that some of the wild yeasts in Oaxaca—like a fast actor named *Kluyveromyces marxianus*—serve as a far better starter for mezcal fermentation than *any* strain of *Saccharomyces* beer yeasts. It not only improves the conversion of fructans to ethanol but retains more potent volatile compounds for better taste and smell.

As for the five dozen or so strains of bacteria found in fermented agave musts, some come along with the agaves themselves, but others reside in the dust in the distillery, the moist crevices in the *canoas*, tanks, and vats, or in the stony surface of the *tahona* mill.

It seems that several different lactic acid bacteria add to the creamy or buttery texture of some mezcals, including three *Lactobacillus* species and two *Weissella* species. These bacteria may reside in the rawhide vats used in Oaxaca, and serve to enhance the propanoic acid of mezcals, giving them a hint of Swiss cheesiness. Others like *Zymomonas mobilis* push up the acidity of the must or *tepache* mash.

❦

To learn what those myriad microbes do to make mezcals such a distinctive realm of the spirit world, enter the labyrinthine laboratories of CIATEJ, a national research center in Zapopan, Jalisco, dedicated to agave products. The center's facilities are staffed by scientists who were trained all around the world to do cutting-edge investigations of why mezcals taste and smell as good and as distinctively as they do. These labs are stocked with the world's most up-to-date liquid and gas chromatography ionization detection equipment, refractometers, potentiometers, next-generation genetic sequencing instruments, and germ-free, temperature-controlled chambers packed with tens of thousands of miniature agave clones in tissue culture.

The labs at CIATEJ and other institutes have determined that 2-phenylethanol (a phenyl alcohol) offers the distinctive floral, rosy, wine-like aromas characteristic of most industrially processed tequilas. As a major product crafted by sugar-loving beer and bread yeasts, it can be saccharine-sweet or bready with floral notes of rosewater or wildflower honey. Some industrial tequilas may also have a waxy or resinous feel to them. Some of the tequila ancestrals and *añejos* have a subdominant odor to them—the smoky, vanilla-like syringaldehyde that is also in coastal *raicillas* and *bacanoras*.

In contrast, another odiferous compound—3-Methyl-1-butanol—is far more potent in the mezcals that are least like *joven tequilas* and more like aged *sake*, the Japanese rice wine. Their volatile aromas can smell like cocoa, caramel, curry, or fenugreek to their aficionados, but have hints of slightly burnt, smoky, sweaty, or tar-like volatiles that sometimes disorient the unacquainted. Once accustomed to them, drinkers are often reminded of an aged cognac or a molasses reduced from quince pulp, with their malt-like nose and syrupy feel. Curiously, some of the stranger flavor volatiles regularly found in artisanal agave spirits, but not in many tequilas—like camphene and benzyl benzoate—are derived not from the plant's genome itself. They are from endophytic fungi hidden within the plant's succulent tissues!

Mezcals from different states and species can have demonstrably different fragrance and flavor profiles. Those from Michoacán tend to be decidedly herbal and fruity, while those from Guerrero have more buttery, vanilla notes and green, "agavaceous" aromas. But wherever the mezcals are from, they tend to have more complex, earthy, medicinal, musty, nutty, smoky fragrances than most tequilas; their flavors are prone to resonate with cedar, oak, and vanilla far more frequently than a sharp, astringent, peppery, immature *tequila joven* does.

Now, here is where all this gets interesting to our taste buds, not just to our brain synapses. These alcohols carry fragrances that would comprise a low percentage of the volume of each batch, were it not for microbes "unlocking" them. Terpenoids like linalool give many mezcals their pronounced floral notes of lemon, lychee, rose, bergamot, blueberry, citron, or orange blossom. To date, limonene (with notes of lemon peel and resinous herbs) and pentyl butanoate (with notes of apricot and pear) have been found only in mezcals produced from the giant maguey pulquero, *Agave salmiana*.

Another volatile fragrance—only recently discovered in agaves—is named farnesol, and it serves in a living agave plant to protect its tissue against the stresses of drought, heat, and oxidation. At the same time, it imparts a mildly sweet, green, herbal fragrance to a mezcal. Recall the citrus burst that you smell when you cut into a Key lime, or the fainter but lingering aroma of a field full of lilies of the valley. That's what farnesol is all about.

You cannot find a phenol called p-cresol in many tequilas anymore, but it is pervasive among ancestral mezcals. Along with guaiacol, p-cresol may heighten the smokiness and barnyard earthiness that is characteristic of many rustic *bacanoras* and *raicillas*. These phenolic volatiles carry ashy, earthy flavors from both the firewood and the slightly burnt *piñas* that are reminiscent of the sweet "geosmin" aromas that you sense when picking up a handful of wet topsoil or kindling from a wood pile just after an autumn rain.

But that is not all. Eugenol offers mezcal a dash of cloves; furfurals a sweet, woody, or bready aroma. Ethyl butanoates sprinkle a mezcal with hints of banana, strawberry, or pineapple. Isobutyric acids impart buttery flavors, notes of vanilla, or the healing aroma of arnica.

We know, we know: That's a strange brew of nerdy chemistry to drink up in one sitting. But, you can bring a clay cup of mezcal up to your nose and simply savor the hints of dozens of dreamy fragrances, some too subtle to even put words to. Then you can tip that same cup and imbibe a distilled beverage that sends shock waves through your mouth, across your tongue, up your nostrils, and down your throat, one wave of flavor after another rippling through your senses.

All of those fantastic flavors and fragrances have been brought to you by often unnamed yeasts and bacteria that make their living by pulling volatiles out of the strange brew of the fermentation vat and bringing them forward for you to taste. Yes, your favorite mezcals are gifts that have come from dozens of different agaves coevolving with thousands of strains of yeasts and bacteria to form mutualistic relationships over thousands of years. But another kind of mutualism is embedded in your drinking gourd as well. Their distinctive profiles of tastes and scents have been heightened by the way they are handled by the human caretakers: women and men—often of Indigenous heritage—who make magic happen in vats, barrel, *canoas*, and skins.

Pedro Jiménez Gurría, owner of *mescaleria* Pare de Sufrir and the Mezonte brand of seventy hyper-local mezcals, reminds us of how much the human touch comes through in the flavors you find in agave distillates elaborated by different *maestros*: "I've tallied at least fifty-five factors that determine the unique flavors of each mezcal, and many of those are influenced by the *mezcaleros* themselves. Each particular producer favors a different set of fragrances, textures, density, or viscosity in his spirits, but each also generates batches that may vary from one another in color, flavor profile, alcohol, or sugar content. A single producer may shift the time of the year for fermentation, its duration, the entourage of yeasts used, the source of water, the kind of firewood, the shape, size, and wood used in the fermentation tanks, and whether the agaves come from recently castrated plants or from ones left to season in the field for a year or two after flowering."

Then Pedro shook his head, took another sip of a *tepe* mezcal from Durango, and laughed: "But that is only part of story. Each person who comes to drink that mezcal—as they bring their own taste memories and even their own genetics of perception up to the bar—receives the flavors and fragrances of the mezcal in an entirely different manner. We will never be able to codify a precise lexicon of flavor in all the agave spirits like the vintners and oenologists have done. The interactions among dozens of agaves and hundreds of human cultures are infinitely more complex than those found in the world of wines. I doubt that we will ever reduce what happens in a fermentation tank down to a simple formula."

To be sure, the fermentation vessels are where the action is; the still itself is simply used to intensify the flavors and to freeze their profile in time. But it is in the wild brew of blessed ferment that the intoxicating sweetness, tanginess, and tastiness of mezcals pulled out of the plant are given a chance to land in your mouth.

Chapter Eight

※

DISTILLING AGAVES DOWN TO THEIR ESSENCE

It is in the final stages in the entire mezcal-making process that the focus turns to distillation, reducing agaves down to their spiritual essence. Regardless of how impressed we are with the huge diversity of agaves used for producing mezcal—as well as the varied traditions of wild-harvesting and cultivation, roasting, crushing, and fermenting—just two or three runs through a still comprise the "spiritual practice" of what *mezcaleros* bring into the world.

The first run of *shishe* (or *común* or *ordinario*) occurs when the fermented mash passes through the still to accumulate the alcohols for greater refinement. The second run or *rectificación* will up the alcohol content and better balance the *puntas, corazón (cuerpo)*, and the *colas* of the various kinds of alcohols with the distinctive aromatic volatiles they carry. At this stage, the *colas* containing toxic methanol and propanol can be eliminated and dumped. An occasional third run can be for finishing and "smoothing out" the aromas and alcohols in the mezcal, or for infusing the mezcal with other flavors, such as those associated with *pechuga* processing.

In short, the answer to the question "What is the distillation of fermented mezcal all about?" is this: the concentration of potency, flavor, and aroma into a liquid that can be stored without spoiling, so that it can be carried in a small container wherever one wishes to take it!

※

The distillers' tasks are accomplished by adding heat to fermented liquids so that the mash or must can be separated into usable fractions. The most aromatic and potent fractions can then be concentrated into storable, delectable spirits. All the other diligent but difficult work of the *maestro mezcalero* is for naught if two or three runs of mezcal through the still do not lead us to the heart of the matter: capturing the quintessential lifeblood of agaves like a genie in a bottle.

※

As if on a spiritual pilgrimage, we rose out of the central valley of Oaxaca into the higher, drier foothills, mesas, and rolling plains of Santa Catarina Minas. It is one of the most celebrated hot spots of innovation in Mezcalandia. In every direction we looked, we could spot patches of different agaves in neat *trazos* or patterned rows, in "messy" but divinely complex *milpas*, or wildly hanging to craggy cliffs above the road.

It was not until we entered the *palenques* that we encountered the special attraction of this small pueblo: the persistence of *olla de barro* (clay pot) distillation, a tradition that has flowered into many significant innovations that are now recognized at a global level.

Passing the fields, the harvested and roasted *piñas,* and the fermentation tanks of the *palenques,* we descended into a shaded area of the distillery where the real alchemy occurs: the row of *olla de barro* stills that generate some of the best ancestral mezcals in all of Mexico. Two separate pot stills often share the same wood fire below and platform above, but the pots at each platform are not connected as they are in many other kinds of stills around the world.

※

First glimpses can be incomplete, but the clay pot stills we first saw in Oaxaca seemed not all that different from the Arabic *al-inbīq* "capped" or "cupped"

Reading of perla *bubbles in* jicarita *to check alcohol content*

stills that we have witnessed in the Middle East. Those are the ones traditionally used for making the distilled anisette known as *araq* or *arak*. In that
ancient tradition, two elongated clay pots (*amphorae*) are connected by a copper serpentine coil immersed in cold water that allows vapors to condense
and precipitate out into another clay pot as an ethanol-dominated distillate.

The roots of terms for anise-infused liquors—*arag, araq, arak, arkhi,* and
raki—appear to have come from an ancient Mongolian term that is a cog-
nate with contemporary Arab and Turkish terms for a still. First used by
Asian alchemists, these terms dispersed across the deserts of Asia to refer
to any alcoholic beverage distilled in Arab-style stills, not only to anisettes. It
is probable that over the last five centuries, at least five different of variations
of Old World stills diffused to Mexico either from Mediterranean ports or
from Asian and Filipino ports.

We can track some of the peregrinations over the last four centuries taken by the most widespread still used for mezcal-making. It is now referred to as the *árabo-mexicano* style of alembic. By the time Phoenicians brought their stills and amphora pots with them to Spain in the seventh and eighth centuries BCE, the "Arabian" style still once known as *al-inbīq* had already become known in Al-Andalus as an *alambique*. This Morisco term then swept into the New World soon after the Spanish Inquisition, when Crypto-Muslims and Crypto-Jews fled from the Iberian Peninsula, first to the Canary Islands, then to the Caribbean islands and the American continents.

The *árabo-mexicano* designs of *alambique* or alembic stills quickly morphed into several forms enduring well into the last century in various regions of Mexico, just as Filipino stills introduced to the coast of Colima persisted in the hinterlands of the Sierra Madre Occidental. It appears that the Filipino/Chinese still design was first introduced through Manila galleon trade to produce liquor from coconuts. It then quickly spread up and down parts of western Mexico, then called Nueva Galicia, to produce liquor from sugarcane and agave.

All this is to say that *árabo-mexicano* variations of *alambique* became the dominant kinds of distillation apparatus used for making mezcals throughout Mexico to this day. Without a doubt, they are not the only, nor the oldest. In fact, at least ten designs of traditional stills have been used in Mexico historically. At most, only six of the still designs currently employed can be clearly traced back to the Mongolian and Arabian roots of the *al-inbīq* prototypes used for anise-flavored, grape-based liquors. The Oaxacan *olla de barro* stills exhibit hints of other, possibly pre-Colonial *(prehispánico)* influences as well.

✳

What we have seen of this still design in Santa Catarina Minas and Sola de Vega remains in a number of other pueblos of Oaxaca today. This style of still may have once been in other Mesoamerican regions in historic times. Instead of teardrop-shaped amphora vessels of clay, it uses sturdy, heat-resistant pots made of a blend of white and black clays glazed with a green *esmalte* slip from Aztompa, Oaxaca.

The *olla de barro* stills are made of two of these stout pots, with the lower one or *caldera* holding 40 to 50 liters of the agave must. In most of Oaxaca, this must or mash is known as *tepache*. It sits in a protective adobe brick casing above an oven packed with firewood, and its open top is parallel with a platform or bench built around it. Fitted on top of the *caldera* or boiling pot is another pot called the *montera* or *capitel*, which has an open bottom so that it functions more like a sleeve. In essence, it is like a clay chimney, cap, or collar that extends above the boiling *tepache* mash so that the vapors rise through it.

The *montera* and *caldera* fit together with a belt of wet clay and agave fiber bagasse that serves as a seal to keep vapor from escaping at their juncture. Fitted into the lip of the upper opening of the *montera* is a copper, iron, aluminum, or hard wooden bowl that serves as a condenser, as cold water is run in and out of it during that boiling of the vapor so that the alcohols and other volatiles condense on the surface of its underside. Surprisingly, Oaxacan campesinos have a particular fondness for the rust-tainted distillates collected in old iron condensers, calling this reddish liquor *mezcal colorado*.

The accumulation of condensed droplets of mezcal alcohols slowly drip into a *palangana* (catchment bowl), a *cuchara* (a hand-hewn wooden spoon), or a trimmed, cup-like agave leaf set at an angle so that the liquor drains out into a *carrizo* cane tube. That tube is set into an aperture in the side of the *montera*, through which the mezcal drains into a funnel-shaped agave leaf.

The 50 liters of *tepache* going through the first run of fractional distillation might take fourteen to sixteen hours to complete. That is slow, tedious work of tending the wood fire and watching that the boiling pot or *caldera* does not burn or crack. Even with pots of the best quality, they seldom last for more than a year, and may break within weeks of setting them in their adobe casing.

Constant repair or replacement of the *ollas de barro* may seem daunting, but they cost a hundredth of what the apparatus in an alembic still costs. Moreover, a modern metal column still of the likes used by industrialists in Tequila, Jalisco, and now by a few in Santiago de Matatlán, Oaxaca, are at least eighty times more time-efficient to produce the same volume of first-run mezcal. But the scant flavors and fragrances emanating from a 250-gallon (1,000-liter) stainless steel column in just four hours cannot hold a candle to those exuding from *olla de barro* stills steaming away for sixteen hours straight. They are not even of the same planet, let alone the same ballpark.

Once we have seen how *olla de barro* stills function, we can understand how different they are from the standard copper serpentine *alambique*, not just from the modern column still. They may not be as efficient at concentrating huge volumes of mezcal liquor in a short period of time, but they work at a much better scale for balancing or rectifying the proportions of the flavorful, fragrant volatiles that have been fractioned into the *puntas, corazones,* and *colas* during the second run. We have even heard of *mezcaleros* so driven for the right balance that they put their distillates through three runs before the final blending.

The *venencia* rectification, or proportional pouring and blending together of three components, is done to raise the alcohol content of mezcal. It may come out of the still with its alcohol content as little as 30 to 50 proof—at least for the first run of *shishe*—to the 70 to 110 proof required to legally call an agave distillate a mezcal. Although some distillers now have refractometers as backup instruments for determining alcohol and Brix content, most have also mastered the art of "reading the pearls" or bubbles of alcohol that rise up in each batch that is passed through a bamboo-like *carrizo* cane called a *venencia*. The *sishe* from the first run typically exhibits little or no *perla* bubbles when a glass or bottle of it is shaken, then allowed to settle.

But from then on, the distiller must give protracted attention to both the size and duration of the bubbles that form as he balances the various components of the spirit. As he blends the *puntas* (180 proof), *cuerpo* (80 to 130 proof), and *colas* (less than 60 proof) of the second run, the *maestro* watches for the largesse, longevity, and consistency of *perlas* that rise up in the mezcal in order to precisely determine the alcohol content of the mixture.

We've witnessed the way Jorge Pérez of Michoacán and Santos Juárez of

Jalisco watch the rising of the pearls while attentively estimating the number of *perlas* relative to those in other batches they have rectified. As they do so, they begins to also pay attention to the aroma and taste of the alcohol to assure themselves that they are in the right ballpark regarding its "proof."

Other *maestros mezcaleros* say that Jorge and Santos are so precise that they should receive master's of science degrees for their skills in reading "tiny bubbles."

Remarkably, a PhD fluid physicist, Roberto Zenit, recently investigated the ancient art practiced by Pérez, Juárez, and others. He determined that reading pearls to guide the *venencia* is not only scientifically valid, but also quite accurate. The Brown University "bubble scholar" decided to change the alcohol content of a mezcal in two directions, first by adding water to one sample to dilute it and second by adding ethyl alcohol to another to make it more potent. In both adulterated samples, the bubbles quickly burst and dissipated.

But in a well-balanced mezcal created through the *venencia* by a master distiller, Zenit noticed that the *perlas* persisted for a longer period because of the viscosity of their aromatic oils. He determined that a good mezcal is also chock-full of chemical compounds called *surfactants* that can reduce surface tension. Surfactants tend to be equally distributed as they linger on a liquid's surface, but when the liquid is shaken and bubbles hit them, many surfactants cling to the margins of the bubble, while others leap to its top, trying to regain their optimal surface tension. As they aggregate atop the dome of the bubble, the bubble bursts.

By glimpsing the "gestalt" of the number of bubbles and their relative duration, *maestros mezcaleros* like Juárez and Pérez are not merely "reading" the alcohol content of a mezcal, but its volatile oil and surfactant composition as well. All these factors truly influence the flavor and fragrance profile of agave spirits.

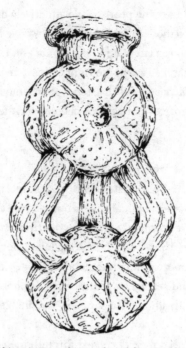

Prehistoric Capacha vessel used for microdistillation

Michoacán ethnobiologist América Delgado-Lemus has documented that the desired pattern of bubble density, duration, and distribution varies with the region and the kind of agave distilled. In southern Jalisco, she notes that *maestros mezcaleros* search for a miniature "pearl necklace" rimming the edge of the container, while in Morelia, distillers hope that larger, longer-lasting bubbles aggregate in a shape akin to a honeycomb. Dr. Delgado argues that these variations in *venencia* balancing contribute to the distinctiveness of each region's mezcal and are part of the intangible heritage of Mesoamerica's gastronomic patrimony.

This stage in the magic of conjuring up "good spirits" is where we see the *maestro mezcalero* as performance artist, one who precisely reenacts the rituals of Asian and Mediterranean alchemists first practiced many millennia before the present. A *maestro* or *maestra* will suddenly whip out two goat's horns or hand-carved *jicarita* cups, then ceremoniously pour the mezcal back and forth between them, astutely observing the rate with which the *per-*

las rise to the surface of the mezcal. Then they announce their assessment of the ABV (alcohol by volume) of the mezcal.

It is at this point in our "spiritual quest" that we might reflect upon how much mezcal making is not only a science, but an ancient art tied to spiritual practices. And only in that context can we best understand the ongoing debate of whether ancestral agave distillates had roots not just in pre-Columbian stills but in pre-Columbian ceremonial practices as well.

Keep in mind that in *olla de barro* distillation, agaves are an integral part of the still itself, in the form of bagasse packed around the juncture of the *caldera* and *montera*, and as the condensation cup and tube siphoning the condensed liquid out of the still. The older *olla de barro* designs hardly used any copper compared to that in the serpentine condensing unit of the *alambique*, for there was little finished metal to be had. A wooden or steel bowl might serve the same function as the steel cap atop the *montera*.

These details suggest that the *olla del barro* still design of Oaxaca was influenced by some other tradition—perhaps a pre-Columbian one. It was one in which local materials such as agave bagasse and trimmed leaves or *pencas* were available at little cost compared to copper and steel. To be sure, these engagements with agave were not just economically or logistically efficient; they probably had ancient symbolic or spiritual dimensions as well.

✺

After the introduction of Asian stills to western Mexico to produce spirits from coconuts or sugarcane, perhaps we reckon that some kind of a mestizo "cultural convergence" of knowledge from the Old World and the New must have enabled agaves to be farmed on a larger scale to produce materia prima for distilled spirits.

This has always struck us as a curious claim—that before Asian or Arabic stills diffused into the New World, Mesoamericans lacked agricultural capacity, technology, and technical know-how to grow agaves to distill into a strong alcoholic beverage. A few archaeologists might begrudgingly acknowledge that Mesoamericans could have figured out the process of distillation, but they made such a minuscule volume of spirits for ritual use that there would be no need for agriculturally producing an abundance of agaves for this purpose alone. By this logic, any pre-Hispanic efforts by Mesoamericans to grow and distill agaves had nothing to do with the *mezcal, comiteco,* and *tequila* beverages we enjoy today.

But around the turn of the millennium, all of those long-standing presumptions were thrown out the window, thanks to a wide array of Mexican scholars like Daniel Zizumbo-Villarreal, Patricia Colunga-GarcíaMarín, Fernando González Zozaya, Mari Carmen Serra Puche, and Miguel Claudio Jiménez Vizcarra. Most of them would agree that they all stand on the shoulders of a rather quiet, enigmatic transborder scholar, Isabel Kelly, who would be a worthy subject of a book all to herself.

In the late 1930s, Kelly and her team of archeologists uncovered an ancient burial site in Colima, western Mexico, where a seemingly revered man had been entombed before the common era. That entombed individual is believed to have been a prominent member of the Capacha culture, a western Mesoamerican society of the Formative period that seemed unlike any other recorded up until that time. The geographic range of those who traded in Capacha material culture spanned much of Mexico's west coast three or four thousand years ago, being found as far north as Baja California and Sinaloa, and southward to Guerrero, if not all the way to Guatemala. Let us just say that the waves of their innovations rippled out over much of the Americas.

Nevertheless, it was the peculiar possessions of this one anonymous man and the funerary items left with him that allowed Isabel Kelly to open a fresh view of the Mesoamerican history of mezcal distillation.

With him in his tomb sat a unique ceramic vessel, with a bulbous base connected by three hollow pipes to a rounded, vase-shaped top. This and another related style of strangely shaped ceramic containers were dubbed Capacha vessels, for the makers of these vessels lived in hubs of Capacha trade networks in Colima, Jalisco, and Michoacán. Kelly was among the first to suggest to other archaeologists that they may have been members of the first complex and technologically sophisticated culture in the Trans-Volcanic Belt of Mesoamerica, one that soon had widespread influence that stretched well beyond their homeland.

Due largely to the lack of research commitment and funding from the Mexican government, Kelly could not initially get the uses of these strange ceramic vessels confirmed. The most curious ceramic vessels were first described as having hollow "stirrups" that connected two superimposed globular vessels that fit together as one. Isabel Kelly so meticulously described and illustrated them in 1939 that her accounts attracted the attention and praise of other cultural historians. The bottom, bowl-like vessel had a bulbous base connected by three hollow pipes than ran up toward a rounded, vase-shaped top. Another bowl could be filled with cold water and fitted into the opening of the larger vessel to serve as a condenser. It closed off any outpouring of fluids, so that they drained out the hollow pipes on the side instead.

While Kelly herself was cautious not to propose in print any special uses of these Capacha vessels, the details of her work prompted speculations by some of the greatest historians of distillation, including the brash but brilliant chemist Joseph Needham. Needham and his colleagues had no qualms at all about suggesting that certain Capacha containers were used for distilling fermented agave juices!

These scholars noted that the vessels had an uncanny similarity to others that were discovered a world away, in Mongolia. The Mongolian vessels were confirmed by carbon dating to be the same age, about 3,500 years old. From the confines of his office at Cambridge, UK, Needham was quick to add that the Asian vessels did have a biochemically confirmed use: some of the earliest known evidence of the distillation of alcohol. More recently, Mexican and French scientists have used new archaeomagnetic techniques to precisely date these unusual gourd-shaped *bule* and three-legged *trifid* Capachan vessels. They too are ancient—3,265 to 3,481 years old.

Then, to the utter amazement of Mexican archaeologists, Needham had the gall to suggest that the hollow "stirrups" of Capacha vessels could be used to perform the microdistillation of alcohol from plants, just as those Mongolian stills in the Old World were used by Asian alchemists. More startling—if not fanciful—was his suggestion in 1980 that Mongolian stills came to western Mexico by means of Chinese navigators several thousand years ago!

American geographers like Henry Bruman initially made mincemeat of the Cambridge professor's hypothesis that distillation traveled across the ocean well before Columbus. Nevertheless, the notoriety of Needham's *Science and Civilisation* masterworks forced other scholars to consider the possibility that American cultures were distilling alcohol long before any Spaniards or Filipino trade ships landed in the Americas.

Needham had simply used Kelly's remarkably detailed descriptions and drawings of Capacha vessels to claim a likeness to Chinese and Mongolian micro-stills. Without having stepped foot in the harbors or mountain hollows of Mexico himself, Needham opened new inquiries of the antiquity of fermentation and distillation in Mesoamerica.

⚜️

So, what if an ancient tradition of agave distillation in western Mesoamerica evolved independently from Asian, African, or European influences? After all, it was simply the two or three stirrup-like tubes running between two ceramic chambers that Needham used to argue for Capacha vessels as fully functional apparatus for distilling agave; Needham himself never offered direct evidence of diffusion from Asia.

Soon, Mexican archaeologists and ethnobotanists on Mari Carmen Serra Puche's team began finding corroborating evidence of the use of mezcal in Mesoamerican roasting kilns, fermentation vats, mini-stills, and associated technologies from the Formative period in western Mesoamerica. When they investigated agave remnants in pottery and in burned soil samples found in cooking kilns in central Mexico, they dated these cultural preparations of agave to two time intervals, between 878 to 693 BCE and 557 to 487 BCE.

They argued not only that pre-Colonial agave distillation in western Mesoamerica was plausible but that their analyses confirmed it. In fact, they suggested it may have been pervasive enough to trigger widespread cultivation and processing of agaves for alcoholic beverages, albeit with extremely small volumes of mezcal rendered per batch.

We should note that some Mesoamerican archaeologists remained skeptical that the unusual shape of the Capacha vessels suited them to be used as distillation apparatus. Others conceded that agave juices could have been distilled into "micro-doses" of a type of mezcal that was probably limited to ritual or ceremonial use, but that had little influence on the more wide-spread historic tradition of *vino de mezcal*. (That, for us, is an untenable argument, for the ritual use of microdistilled mezcal would likely have as much cultural and spiritual significance as small sips of communion wine or peyote in other cultures!)

Still other archaeologists said the proof is in the pudding—or the pulque. They insisted you must find the unique plant chemicals of agave embedded in the ancient ceramics to have forensic evidence that confirmed Needham's flamboyant (or preposterous) proposal.

Almost two decades passed before Mexican scientists came up with inge-nious ways of definitively testing Needham's 1980 hypothesis. After first confirming that agricultural production and agave processing technologies flourished in Colima during prehistoric eras, Daniel Zizumbo-Villarreal, Patricia Colunga-GarcíaMarín, and Fernando González Zozaya did a simple experiment to quell the skeptics. They used a Capacha-style gourd-shaped *bule* vessel to successfully distill a fermented agave juice into a small quan-tity of mezcal with 20 percent alcohol, and then a three-footed *trifid* vessel to distill another batch with 32 percent alcohol. The assumption that Meso-americans were somehow incapable of domesticating, agriculturally produc-ing, and distilling mezcal before the Conquest fell with a thundering crash!

When we visited Daniel and Patricia at their home in Merida, some years later, they were gracious enough to show us a few of the replica Capacha-style vessels of the kind that had been used to distill the fermented musts from cultivated agaves in western Mexico a couple thousand years before us. We stood there amazed, as if we were seeing a Rosetta stone that explained how *jimadores* and *maestros mezcaleros* had begun to shape the agricultural and gustatory traditions upon which the mezcal industries of Mesoamerica and Arid America were founded twenty-five centuries before us.

Another clay vessel used for mezcal distillation came into the hands of researchers at Temple University for evaluation. Employing a 7 Tesla electromagnet the size of a Mini Cooper, Drs. Swati Nagar and Ken Korzekwa are identifying traces of compounds found only in agave. They will attempt to identify the patterns of oxidation specific to distillation, and perhaps then confirm agave spirit distillation in ancient samples. This analysis comes after years of academic drama and interpersonal conflict between US and Mexican researchers on whether the quality of evidence fully confirms or denies that there are biochemical residues of distillation to be found in ancient Mesoamerican ceramics.

Now then, what real difference does the antiquity of Mesoamerican distillation really make to contemporary drinkers of mezcal? Will it make each mezcal taste richer if we can deeply muse upon its origins? Well, in one way, yes. Just as psychologists claim that the largest sexual organ that humans must employ in lovemaking is the imaginative mind, it is also true that our imaginations influence what we taste in a dreamy mezcal.

If we believe that the mezcal we hold in our hand and sip into our mouth is the tangible evidence of a great human achievement that occurred in western Mesoamerica 2,500 to 3,500 years ago, that wondrous event fills up our glasses and our minds. It influences our taste buds. It gives us an epic story to tell our friends as we sit by a campfire and imbibe our mezcal. Soon enough, the world will know without a doubt whether Mesoamericans were using it for ceremonial agave spirits distillation thousands of years before the Spanish, Filipinos, or Chinese supposedly taught them how.

When this debate about the possible New World origins of distillation is settled—whether the consensus finally and firmly recognizes that ancient Mesoamericans distilled agave spirits long before Spaniards, Filipinos, and Chinese came ashore in the Americas—we may have more to celebrate. We can properly express gratitude for the diversity of human innovations that contribute to the many pleasurable terroirs of agave spirits. And when an abundance of gratitude is present among friends who come together to share their favorite spirits, it always makes the mezcal taste better.

Chapter Nine

❧

THE TEQUILA PANDEMIC
AND ITS DISCONTENTS

As much as we love the smoky fragrance of some mezcals, we would be deceiving you if we did not concede that the smell of danger is in the air. In David's home state of Jalisco, we have witnessed the horrific consequences of multinational conglomerates diluting the potent heritage of different agave spirits and pouring money into efforts to convince regulators to compromise the historic standards of integrity.

We are surely not the only ones who have come to this conclusion. We have heard the same concern raised from growers, from agronomists, geneticists, and microbiologists at technological institutes and the National University (UNAM) rigorously trained in objectivity, and of course from owners of small- and medium-scale distilleries themselves.

Over the last three decades, we have all watched listlessly as the freshest, fieriest agave spirits were hosed down to make them "smoother" for more moronic versions of cheap margaritas. If you start off with crap, no cosmetic surgery or air freshener is going to make it look more handsome or smell any sweeter. And as one well-regarded agronomist inside the tequila industry confided in us, the so-called tequila boom has generated disastrous consequences in the field in terms of soil health, the price of planting stock, the genetic vulnerability of agaves to insect pests and plagues, and also the health and welfare of the laborers in the fields and distilleries.

Just why did the Denomination of Origin for tequila go south, failing to protect both the integrity and distinctiveness of the world's best-known agave distillate? In the post-NAFTA era, how did the neoliberal globalizing pressures on the beverage industry force its transnational corporations to market tequila *mixto* as if it was the Bud Lite of agave distillates?

Let us give you a brief autopsy, a look into the forensic evidence of where tequila went down the wrong road and crashed. It is much like unpacking a bad accident, so you will have to be patient with us for a moment as we descend into the darkness before returning to full daylight.

As tequila began to garner recognition across Mexico in the mid-1800s, distilleries were concentrated on the haciendas of some of Mexico's most powerful families. These wealthy *hacendados* pushed hard against the spirits' Indigenous roots to elevate their own social status. Mexico's classist society had always dismissed its Indigenous peoples, their foods and beverages, as being of marginal value to the world at large. It followed that if the Mexican elite were to maximize the social and economic capital that agave spirits could provide them, they would have to wipe them clean of their Indigenous connections and nativistic connotations.

Thus began the construction and fierce protection of the myths about European viceroy Pedro Sánchez de Tagle, who falsely received credit for having introduced *vino mezcal de Tequila* to the world. (When uncovered documents showed that the viceroy was nowhere near Tequila, the documents were placed in the archives of the Museo de Tequila, where they were soon targeted in a Molotov cocktail attack. The archives survived the fire, and there remains the proof that this manufactured history of Tequila is one part science fiction and another part *caca de vaca*.)

As the most powerful *hacendados* invested more of their efforts in controlling the destiny of *vino mezcal de Tequila*, they found themselves uniquely positioned for expansive success. The Jalisco elite, such as the Sauzas and Cuervos, married their daughters off to German engineers, who brought in stainless steel tanks and modern technologies to replace more rustic equipment, so they could make a "cleaner" tequila. Opal, silver, and gold mines drew labor to just south of the region, and masses of underpaid miners created demand for alcohol. As Mexico's mineral wealth and precious metals were extracted to make their way north into the western United States, railroad shipments passed through the town of Tequila, Jalisco, where crates of bottled alcohol were added to the freight.

As more of the region's spirits found their way north across the border, demand for this mysterious liquor soared in the US, and convenient, fatally flawed, and fabricated myths of tequila's origins went along for the ride. With empire builder E. H. Harriman's investment in the Southern Pacific Railway from Mexico to the US, railcars loaded with tequila began to arrive in bigger cities and towns of the Wild West just prior to the Mexican Revolution. There was already enough civil unrest on both sides of the border to multiply the number of thirsty soldiers of fortune in the region who would seek out the hard stuff and "hooch" of any quality.

But to the ragtag troops on the other side of the border, this Mexican moonshine was not a sacramental spirit of Indians; it was the spirit of royals and merchants. By 1901, Mexican distilleries produced nearly 10 million liters of agave spirits, much of it destined for the border. That meant that mercenaries and vigilantes in the borderlands were requiring the harvesting, roasting, fermenting, and distilling of more than seventy million agaves per year.

Riches compounded and demand was sky-high, but there was one glaring, lingering problem staring at the Sauza family: The raw material for these spirits, the majestic agave known as *el árbol de las maravillas*, seemed to take longer than a towering tree to render its marvels. Even after fifteen to thirty years of growth in drier climes, most of the agave varieties harvested for distilled spirits hardly had enough convertible carbohydrates to produce as much alcohol as you could get by raising a small patch of cane for its sugars.

That is, with the exception of just one peculiar, cultivated variety or "cultivar" for short: the rather precocious, blue-hued tequila variety. Although it was proposed nearly a half century earlier by the French botanist Frédéric Albert Constantin Weber, the scientific name *Agave tequilana* Weber cv., *tequilana azul* was formally accepted as a new scientific name in 1902.

At this pivotal moment in the economic history of agave spirits, the bluesy mutant was reclassified and elevated to the level of a completely distinct species. Ironically, Weber spent only three years of his life in Mexico (1864 to 1867), where he was gainfully employed not as a botanist but as a physician for the French military.

That is a mighty short period for a botanist to identify a potentially unusual plant, to tentatively describe it, and then to compare its morphological distinctions and geographic limits with other species found in Mexico, even if that botanist is working at it full time. We can assume that most Mexican *maestros* and taxonomists have needed many more observations in order to discern whether an agave like the *tequilana azul* cultivar is distinctive enough to be named a unique variety or species.

Not long after the turn of the millennium, Gary worked with botanist Ana Valenzuela to show that Weber's famed agave is just one more variant of the most widespread species of agave in the Americas, *Agave angustifolia,* but our taxonomic change was not accepted, especially by the tequila industry or nomenclatural gatekeepers. And yet, most prominent Mexican agave scientists have independently come to the same conclusion.

Today, most marketers of tequila unnecessarily include Weber's name in all their propaganda as if its association with a Frenchman legitimizes or reifies it being set apart from all other agaves. That is ironic, for in common parlance, few botanists ever recite the surname of any other taxonomic authority as part of any other agave's scientific moniker. It is not only unnecessary but rather silly to conjure up the ghost of the bourgeois doctor every time one says *Agave tequilana*! It has become yet another classist convention of the tequila industry.

This cultivated species, now legally known as *Agave tequilana* Weber—

epitomized by a single cultivar, *tequilana azul*—was then reaching sexual maturity and harvestability in just ten to fourteen years, and could reproduce by cloning itself. Its *pencas* are packed to their gills with sweet fructans. However, as the climate of tequila-growing regions became hotter and drier since 2002, plants propagated from vegetative offshoots began to reach maturity in nine years, then in just seven years. Today, vegetative offshoots propagated in nurseries and then outplanted into fields are being harvested in as little as five and a half years, in immature condition. Worse yet, clonal plantlets tissue-cultured in petri dishes or test tubes are being harvested in under three and a half years from outplanting. By 2020, these tissue-cultured clones were being harvested in half that time, mature or not.

As CIATEJ scientists have determined through their own empirical studies, it is far more economic (and delicious) to let *tequilana azul* plants grow exponentially as they mature in fields for a rock-bottom minimum of seven years, rather than harvesting immature heads in half that time. You gain more kilos of edible biomass over two decades of time, require less hand labor for trimming and weeding, use less herbicides and pesticides, and suffer less loss of soil microbial health. A few more years in the ground makes them yield heavier heads and become even more flavorful.

But what does common sense have to do with growing *tequilana azul* in this day and age?

Aside from the devastating effects of using immature plants on the spirits' quality, this greed-driven move created an expansive monoculture we call the "blue desert." It now stains entire landscapes blue throughout large swaths of Jalisco, but also in Sinaloa, Nayarit, Tamaulipas, and other states where nursery stock is being propagated before its return to Jalisco. There are now six hundred million *tequilana azul* plants being grown on more than 700 square miles (180,000 hectares) within certain municipalities in Mexican states.

Several of the larger producers then made a Faustian bargain to keep up with the global demand for their premixed margaritas and their tequila shooters. They built clandestine, highly guarded laboratories to "tissue-culture" clones of agaves in petri dishes and test tubes. At first, the labs were hidden from public view and access to other scientists was strictly prohibited, because the federal government had a ban on genetically engineering crops that were considered part of Mexico's patrimony. As reported by Rex Dalton in *Nature* in 2005, it remained hotly debated whether some tissue culture techniques were legally prohibited or loosely permitted under the federal ruling. But it was clear even during that era that some labs were considering—if not initiating—genetic selection techniques through other biotechnologies. These lab propagation processes could potentially multiply, speed up, and select the number of improved *tequilana azul* clonal offshoots to be available for outplanting in the fields.

In 1989, while working on a traveling exhibit about tequila and its history, Gary was unexpectedly invited to see one of these tissue culture labs run by the Mexican scientists working for one of the industry's largest distilleries. It was stunning to see how sophisticated the lab was being managed on a budget that seemed lavish compared to that for the laborers' activities in humble fields surrounding it. That was when the federal government was shaping its ban on genetically engineering crops that were considered part of Mexico's patrimony. At the same time, the well-trained horticulturists in their white lab coats aspired to engineer ever-faster-maturing agaves that could be free of both diseases and the need for "messy" sexual reproduction.

To this day, Mexican scientists continue to debate whether simply cloning and micro-selecting of "superior" agaves by tissue culture in petri dishes should be legal in their nation. However, CIATEJ scientists in Guadalajara did take us up to the glass doors on a climate-controlled tissue-culture repository to show us genetic engineering of a clone of *Agave tequilana* that has resistance to the cocktail of agave diseases commonly known as *tristeza y muerte*.

We were confounded that such a technological feat had already been accomplished, while neither of us had ever even heard of the selection being grown in the field for tequila production. The CIATEJ staff quietly noted that the Tequila Regulatory Council (CRT) had requested the suspension of its release, for it was unsure whether it would be legally considered a "strain" or "variety" of agave that was genetically different from the disease-susceptible *tequilana azul* clone that their entire industry had been built around.

The disease-resistant wonder has remained in mothballs to this day, despite the continuing spread of *tristeza y muerte* to other regions and to other wild agave species.

To date, these pilot projects have not yet released or mass-produced transgenic cultivars with herbicide resistance like those conjured up in biotech laboratories for corn and cotton. But they did open the door to growing hundreds of thousands of genetically identical agaves in vitro under tissue culture. That innovation was not at all a good step toward maintaining a diverse genetic base for the tequila industry.

Without a doubt, the blue tequila plant was never the most complex, delectable, or flavorful member of the forty-two species in the agave family being used to make mezcal in twenty-four Mexican states. Nonetheless, this rather anomalous upstart was efficient enough to meet the growing demand for "Mexican moonshine" in the US. Its sales north of the border (to drinkers who knew little of other mezcals) soon eclipsed the traditional demand in Mexico. South of the border, pulque from fermented agave sap remained far more popular than tequila for at least five more decades after *tequilana azul* was formally named.

Certainly, novice consumers of agave spirits in the US were not accustomed to the complexity of most mezcals, or even the multi-varietal tequilas at that time. In the same historic moment, Mexico's elite could import and drink spirits from Europe; they were happy to distance themselves from the traditional beverages of Indigenous cultures. They associated mezcals, pulques, *tesguinos, tepaches,* and *colonches* with the lowest rungs of their society.

In fact, for most of the twentieth century, Mexico's elite consumers flaunted their preference for "classier" imported spirits like whiskey over any agave spirit. Indigenous connections from Mexico's rustic past were a hindrance for promoting agave spirits among urban consumers and producers alike, and over time, they faded from memory.

By the 1940s, the wild success of the haciendas bestowed all the industry's power onto just a few families, who used their political capital to sway federal laws in an effort to secure efficient if not more exclusive production of mezcal's prodigal son, tequila. As young American men returned home from World War II, demand for hard liquor and sweet cocktails swelled in the US, so much so that Mexico's *tequileros* were keen to supply them by any means necessary. From the late '40s onward, the industry grew consistently more industrial.

As the haciendas honed their focus on filling bottles quickly and at low cost, they successfully lobbied for a change that would haunt tequila's future: added sugars. For the next thirty years, a spirit with up to 30 percent non-agave sugars could be sold as tequila. After an unfortunate adjustment in the 1970s, the allowance increased and remains today at 49 percent non-agave sugars. Corn and sugarcane syrups have become the favored substitutes, and *tequila mixto* has become largely synonymous with splitting headaches and vomiting into bedside trash bins.

In effect, various watered-down regulations damaged the spirit's historic complexity, which led to a leveling-off in quality. No smaller producer could enter the scene with a more nuanced traditional spirit and market it with the hottest word emanating from south of the Río Grande: *Tequila!*

Suddenly, tequila had become a household word in the US, blaring off transistor radios with versions of a pop hit by that name covered by the Champs, the Ventures, Dizzy Gillespie, and Boots Randolph. Market share expansion rather than quality would become the be-all, end-all indicator of success, and by cutting out producers who used other agaves, the *hacendados* captured even more of the share.

At about this time, Gary teamed up with Ana Valenzuela to sound the first global alarm about tequila monoculture in a series of articles, lectures, and a book—*Tequila: A Natural and Cultural History*. When we visited the Tequila Chamber of Commerce and alerted the Tequila Regulatory Council of the widespread concerns among scientists regarding the trajectory of their industry, they were not amused. The top brass in the chamber were religiously convinced that new agrichemicals could and would take care of any pest or disease that emerged in the blue deserts of their *tequilana azul* clone.

That is both tragic and ironic, because it slowed down the search for other options to deal with the genetic vulnerability of their monoculture. Today, CIATEJ plant pathologists have demonstrated that microbial inoculation of the roots of field-planted agaves provides the crop with stronger, more cost-effective protection against most diseases than costly agrichemicals can offer.

If the CRT had shifted its focus from thinking that an agave clone lived in isolation, toward viewing it as a *holobiont* whose health was as dependent on beneficial microbes and pollinators as it was on its own genes or on pesticides, this soap opera might be over. But it is not.

✺

Not long after our perfunctory meeting with what growers now call "the cartel of power" in the tequila industry, Dr. Katia Gil-Vega announced that she and her colleagues had scientifically confirmed beyond any doubt the extreme genetic vulnerability of *tequilana azul* clones: "This is one of the lowest levels of polymorphism detected to date in the analysis of a crop species and is proposed to be the result of the promotion of a single conserved genotype over many years due to an exclusive reliance on vegetative propagation for the production of new planting materials."

In essence, the *tequilana azul* cultivar had become as badly inbred as the Spanish Habsburg kings and queens whose litters of princes and princesses suffered from protruding jaws, twisted tongues, hemophilia, and hydrocephalus.

Prominent Mexican scientists in several institutions soon agreed with our prediction that because blue tequila agaves had the most severely reduced levels of genetic variation ever documented in any commercial crop, they were becoming increasingly vulnerable to catastrophic diseases. In 2012, researcher Miriam Díaz-Martínez and her colleagues added another nail to

Agronomist pulling up agaves infected by tristeza y muerte *diseases*

the top of the coffin with regard to the loss of long-term viability in *tequilana azul* monocultures: "Low levels of variation were observed between mother plants and offsets. . . . Families obtained from commercial plantations showed lower levels of variation in comparison to families grown as ornamentals. No variation was observed between the original explant and four generations of in vitro cultured plants."

Today, many agave experts agree that the genetic variability left in *potrero* plantations of *tequilana azul* has fallen below 73 percent, an extreme level of impoverishment compared to the rich heterogeneity found in most other agave species, wild or domesticated.

Tequila has become the equivalent of going to a Baskin-Robbins ice cream shop to savor its thirty-one original flavors, only to find just eight or fewer kinds of ice cream in the bins. It is like going to the Fabulous Flower World superstore to find just one floribunda rose left in the pots.

Although *tequilana azul* clones were still peaking in their production by 1996—when two hundred million vegetatively propagated plants were placed in the fields—there were already unmistakable signs that something had gone badly amiss. As higher temperatures and wetter weather enveloped most tequila-growing regions, the *jimadores* or harvesters started to complain about the *marchitez*—a "withered condition" affecting nearly a fourth of the agave plants under their care.

Our colleague Ana Valenzuela painfully described the arrival of a new "cocktail of diseases" in a 1998 report from the University of Guadalajara: "Even in formerly healthy plants, the symptoms of wilting and withering have suddenly appeared, beginning with discoloration of the agaves' leaves. Days later, it is apparent in all the plants' tissues, from roots to shoot. Within months, the plant is found to be completely devastated, desiccated and weakened to the point that a little nudge is enough to topple it to the ground."

By the end of 1998, Ana estimated that nearly 235 of the *potrero* plantations of *tequilana azul* spot-checked in her region suffered from symptoms of one or more variants of the disease. Suddenly, the trouble in tequila fields was being likened to the Irish potato blight that triggered a famine in 1845. It was also compared to the Southern corn leaf blight epidemic that hit the southeastern and central US in 1970. The narrow genetic base of most hybrid corns in the US contributed to a billion-dollar loss to farmers, as yields dropped 20 to 30 percent in just one year. Now, tequila was speeding down the same dead-end road.

Most of Mexico's plant pathologists were baffled by this tequila pandemic, for they were unable to link it to a single pathogenic microbe. When they sampled the pathogens found in the withering agave plants, they identified five distinct disease organisms that appeared to be working synergistically to take down most vulnerable agaves within their reach.

Early on, they identified a fungus named *Fusarium oxysporum*, which they believed had triggered the rotting of the meristem of hundreds of thousands of blue tequila agaves. But a few years later, more determined pathologists identified another, lesser-known fungus, *Fusarium solani*, as the likely trigger for a reddish-colored root rot found in the diseased *tequilana azul* agaves and for the plants' loss of capacity to anchor themselves in the soil.

It didn't take long for yet more teams of scientists to spot three additional strains of bacteria that were also contributing to the soft rot in the genetically uniform agaves: *Erwinia cacticida, Pantoea agglomerans,* and an unidentified *Pseudomonas* species. Still another pathologist raised the possibility that a rod-shaped, gram-negative bacteria, *Enterobacter,* was also contributing to the fall of the towering tequila plants.

Now, here is the punchline: No agrichemical miracle could relieve *tequilana azul* agaves of their multifaceted malady. There was simply no quick chemical fix nor agronomic solution to a multiple-disease pandemic triggered by the extreme genetic vulnerability of agaves fostered by the Denomination of Origin restrictions. This wicked brew of six distinctive pathogens combined forces in pairs or triads in each field to take down roughly 200 million *tequilana azul* plants by 1998, with a quarter of the crop infected.

The tequila industry began to stagger around as if it had a bad hangover, while many of its blue plantations went into a terminal coma. The acreage planted to the *tequilana azul* clones plummeted by a quarter, as dead and dying plants were pulled from the ground to stop the spread of these diseases.

As if things couldn't get worse, a rare early freeze then killed off a good portion of the immature agave plants that growers had hoped would replace the more mature ones that had already succumbed to disease. There was little left to work with to ensure sizable harvests over the next few years.

In June 1999, Ana Valenzuela offered the *Boston Globe* a dire prognosis: "It's hard for us to admit that our national product is in danger, but this problem is getting worse. . . . I don't know what the small producers will do."

To be sure, the small-scale *tequileros* did not know what to do. The owners of at least thirty-five distilleries gave up the ghost and went out of business. A month after Ana's pronouncement, Mexican economists projected that the lost revenue from these diseases on 150,000 acres would amount to nearly $200 million. With a loss of sixty-four million plants needed as materia prima for tequila production, twenty more distilleries closed shop by the summer of 2000.

The costly consequences of overreliance on a genetically vulnerable monoculture sent shock waves throughout the beverage industry that ripple to this day. In fact, we have recently heard reports that the *tristeza y muerte* cocktail of pathogens has now jumped from newly expanded tequila fields to adjacent mezcal fields and wild agaves within formerly disease-free municipalities in southern Jalisco. Some worry that it is likely to arrive in the *espadín* monocultures of Oaxaca as well.

Fortunately, a phoenix has risen out of the ashes of the tequila industry. Mexican producers and their angel investors decided not to put all their eggs back into the tequila basket, but instead to hedge their bets. Today, more than three hundred distinct agave distillates (including mezcals) are produced from over a hundred named agave varieties. They are sold into sixty-eight countries in the world market. Some restaurants and bars in the United States, Europe, and Japan now feature eighty different brands of mezcal that were virtually unknown when the great genetic collapse of tequila began.

Although mezcals still garner only 2 to 4 percent of the market share of all agave spirits (including tequilas), they have been the fast-growing component of Mexico's beverage industry for most of the last two decades. By 2016, mezcal exports surpassed domestic consumption in Mexico for the first time. That year, producers sold a record volume of mezcal: 640,000 cases both inside and outside Mexico.

The mezcal market was valued at $727.11 million in 2019. Because mezcal sales worldwide are projected to reach $1,136 million by 2027, the growth rate could be as high as 6.1 percent from 2020 to 2027. It is far more difficult to summarize the growth of demand for 100 percent agave distillates other than mezcal, for reasons we will later explain.

The exports of mezcals to the US have been rising by 23 to 27 percent year after year, and now account for a fifth of all agave spirits exports. US consumers now sip more imported mezcal than Mexicans enjoy on their home turf. Accordingly, American bartenders and consumers are recognized by the Mezcal Regulatory Council as relevant stakeholders. They—you, us— are now offered places at the table in decision-making regarding the future of the agave spirit world.

There are good reasons to believe that by working together with producers and bartenders, we can build a healthier, more diverse future for all agave spirits produced in Mexico than the one now put forward by the dominant players in the tequila industry. An unprecedented diversification of distilled agave spirits as well as a flush of new agave-based food products are becoming so well rooted in the Mexican and global economies that it is unlikely that they will ever vanish.

The current attractiveness of many mezcals threatens to place some of them in harm's way, especially as global demand is set to increase threefold by 2027. That is why we must now face those challenges and solve them, rather than dismissing them as most of the *tequileros* chose to do during their moment of Big Boom before their Big Bust.

After eight thousand years of agaves being cared for, tended, and consumed in Mexico, Guatemala, and the US Southwest, it seems as though they are on the verge of recapturing their rightful places in our food and beverage economies.

PART TWO

THE FUTURE OF THE HUMAN-AGAVE SYMBIOSIS

Chapter Ten

FROM STILL TO BARSTOOL

Agave distillates have traveled a long way over the last century and a half, when they broke out across the Mexican borders to parts unknown. Today, no fewer than 250 brands of tequila have legally crossed the borders of Mexico to travel overseas, carrying their original "passports" with them. That is nearly a fifth of the 1,300 registered brands of tequila produced by 150 distilleries in the country. But to our astonishment, there are just as many mezcals that make it out from the distilleries to dazzle strangers in foreign lands, with well over 300 mezcal brands registered in Mexico now exported. That is twice as many brands as there were certified by the Mezcal Regulatory Council in 2015, a doubling that occurred in just seven years.

Nearly all of the brands of mezcal have submitted legal paperwork for export permits, given that mezcal consumption outside the country is far greater than within the republic. Already, another two dozen brands of *raicilla* and *bacanora,* are exported, but it is more difficult to track the travels of these spirits. Only Mayahuel the Agave Goddess knows how many 100 percent agave distillates are savored in countries other than Mexico.

But here's the kicker: Tequilas and mezcals also go for higher prices per liter outside Mexico than within. Tequilas touted as *Clase Azul Extra Añejo Ultra* regularly go for as much as $1,700 a bottle in Europe, Japan, and the

US, while an elegantly seasoned *pechuga* mezcal from Oaxaca brings more than $200 a bottle.

Of course, considerations other than age, quality, and distance traveled influence ultimate going price. We hear more about the absurd anomalies, like the liter bottle of Aztec Passion Limited Edition from Tequila Ley 925 that sold for $225,000 in 2006. It was registered by Guinness World Records as the most expensive bottle of agave spirits ever sold.

That sum may be due to the fact that the tequila was encased in 4.4 pounds of gold and platinum. Tequila Ley's CEO Fernando Altamirano insisted, "This is a really unique bottle of tequila and our client, a US-based collector of fine wines and spirits, will treasure this prize to add to an already impressive collection," without disclosing the buyer's name, IQ, or blood alcohol content at the moment of purchase.

On the other hand, some of the most memorable mezcals and agave distillates ever produced sell for little more than $75 to $100 a bottle overseas, and for as little as $35 to $50 directly from the *maestros mezcaleros* at hometown *mezcalerías*.

Understanding each link in the supply chain from still to barstool can help explain where in the chain the price per bottle changes most drastically. For every liter bottle of mezcal that sells for 1,000 pesos in a Mexican liquor store, at least 670 pesos (averaging 69 percent) of that price goes to cover two taxes—the Special Tax on Production and Services (IEPS) and the Value-Added Tax (IVA). The first tax helps the Mexican government deal with the environmental and social costs of production, use, and abuse of distilled alcohol, tobacco, sweets, syrups, and so-called energy drinks, as well as fat- and fructose-laden fast foods. The second applies to the sales of all commodities, but with Mexico's unrivaled bureaucratic flourish, it essentially functions as a tax upon a tax when it is applied to agave spirits.

If the federal government takes almost seven-tenths of the street value of every bottle of mezcal that is formally registered with Mexican agencies, you begin to see why there are incentives to export or to clandestinely sell agave spirits into the black market. The populist preference for bootleg mezcals and other spirits is a form of resistance against the bureaucracy that has insinuated itself into the beverage industry. Without a doubt, buying bootleg remains a necessity for many campesinos who still live on a tight budget. And while selling mezcal into the global market allows *mezcaleros* to avert paying the high costs of IEPS and IVA in Mexico, they must nevertheless "pay to play" to position their product in the most lucrative export markets.

In ways, it can be a lose-lose proposition for small-batch producers. They must pay out the wazoo even to legally offer their distilled juices to their neighbors, or they must pay thousands of dollars for paperwork before their first legal bottle leaves the republic. Neither option is satisfying.

David recalls the moment when he and Salvador Rosales Torres prepared to send off Lot #1 of Siembra Valles Ancestral Tequila Blanco to cross the Mexican border with Texas. The Consejo Regulador del Tequila (CRT) had approved the product and its label, but an inspector noticed that the word *mezcal* was mistakenly used in the description on the label. That one word meant that it must be approved by the Consejo Regulador del Mezcal (CRM) as well, but the same spirits could not fall under both regulatory council's auspices at the same time. As a result of this bureaucratic snag over one word on the label, Salvador and David had to reprint labels that would be sent up to the border in future shipments.

Perhaps no distillates in the world voluntarily carry such detailed eco-labeling as do mezcals and their kindred spirits. Through all the details offered on the label, you come to know and respect the *maestro mezcalero* and family, and the cultural landscape in which they live. You learn something of the soil and stone, the spring water or stream that underlie their operation and influence a particular terroir. You learn of the woods they use in kilns and ovens, in fermentation vats and barrels. You learn the kind of stills they use, from copper-coiled *alambiques* to *ollas de barro*. You learn agave species and varieties used, and whether they were castrated before flowering or left to season for up to four years after seeds were set. On some labels, you can find out whether bats were allowed to pollinate the blossoms, and whether reforestation efforts are built into the price of your product.

To invest in such practices takes time, and often considerable cash. For instance, the first four organic mezcals sanctioned for export into the

US—Del Maguey, Montelobos, Mezcales de Leyenda, and Wahaka—all had to jump through the hoops of USDA Organic Certification, over and beyond Mexico's Compliance Program (MCP) for certification to the Mexican organic standards (LPO). USDA Organic branding often costs as much as $5,000 per product to finesse. And yet, Raza Zaidi of Wahaka sounded unfazed by it all, claiming that USDA certification went through without complication because "we were always 100 percent organic by tradition." According to Zaidi, only organically grown, pesticide-free agaves, clean water, and non-genetically engineered yeasts were used by Wahaka, just as Oaxacans have done for centuries.

And yet, such scrutiny can either amuse or frustrate the many Indigenous mezcal producers in Oaxaca and Puebla, who in effect must prove their own Indigenousness—by filling out paperwork in the colonial languages they often do not speak on a daily basis.

Once agave spirits leave the distillery, they then pass through the hands of many other agents who will each get their piece of the action—their own cut of the retail price on the street or in the bar.

Of course, not all agave spirits leave the *vinata, taverna,* or *palenque* along the same routes, nor do they follow the same protocols for export. Some western Mexican mezcals and tequilas are shipped by boat southward down the Pacific coast, cross through the Panama Canal, then head southeast to Cartagena, Columbia, before turning north to reach Caribbean and Atlantic ports. Leaving the port in Manzanillo, Jalisco, and passing through the Canal, tequila and mezcal are shipped 4,200 nautical miles over eighteen days at sea to arrive in one of the ports surrounding New York City.

As far as passport data go, the regulatory councils for *mezcal, raicilla, bacanora,* and *comiteco* often follow the protocols first shaped by tequila regulators, but some aspects of categorization are not comparable from one spirit to the next. For instance, mezcals are categorized by three degrees of aging—*mezcal reposado* (aged in wood containers for two to twelve months), *mezcal añejo* (aged for more than a year), and *mezcal madurado en vidrio* (aged in glass containers underground for over a year)—but there is no *extra añejo* category as there is for tequila. The spirited juice sold as 100 percent agave distillate follows no such set of formally defined categories but can use such terms in a more informal manner.

Keep in mind that the unaged mezcals referred to as *joven* remain the preferred drinks in bars and restaurants around the world, followed by the *reposados*. In most cases, *mezcal joven* brands are cheaper than other types, because of the lower production costs of aging them two months or less . . . or not at all. They are often crisper, sharper, or more peppery, with white pepper, wood smoke, citrus, and green apple taste notes often dominating.

Some of the most well-traveled *mezcales jovenes* come from the increasingly industrialized *palenques* around Santiago Matatlán, Oaxaca. That is what you probably get in most mezcal cocktails. But sipping *añejos* and *reposados* straight up has become common among the cosmopolitan hipsters who are willing to pay extra for the varietals marketed by fast-growing, midsized brands.

To be sure, unusual agave spirits have garnered much more attention by aficionados and journalists recently than they received in the last millennium. Nevertheless, the richness of the more obscure *bingarrotes, chichihualcos, huitzilas, jarcias, quitupanes, raicillas, tasequi, torrecillas, tuxcas,* and *zihuaquios* remains altogether unfamiliar to most thirsty citizens on Planet Desert.

In this moment, the Denominations of Origin—especially that for mezcal—offer umbrellas that shelter fewer than half of these historic distillates of agaves. To us, that reveals that a broad suite of agave spirits is simply being ignored or even suppressed. They do not travel very far from their distilleries, and seldom if ever reach the bright lights of bars in faraway cities such as Puyang, Paris, Perth, Palermo, Porto Alegre, or Pretoria.

Many producers of the distinctive agave distillates wish to remain closer to their humble agrarian roots rather than "shooting for the moon." They are reluctant to place the destiny of their best products in the hands of execs in high-end marketing firms, who bump up the price of the products 20 to 25 percent for merely visiting a *vinata* twice a year. And so, a growing number of distillers simply call their products 100 percent agave distillates, selling them locally in cantinas in bulk *(al granel)*. Marketing distillates of 100 percent agave is already sanctioned in Colima, Chiapas, Jalisco, Nayarit, and Sonora. In Sonora, Chihuahua, and Durango, young distillers continue to combine agave spirits with those of sotol *(Dasylirion)* just as their bootlegging grandparents did, in blatant defiance of the DOs for *bacanora,* mezcal, and sotol.

At least for now, these and other small-batch distillers will continue to bootleg their spirits, remaining below the radar of any regulatory council. Although the Mexican government does not record data about them, moonshined "mezcals" continue in clandestine production in at least twenty states, and perhaps in as many as twenty-four states. Indeed, the contributions in volume of bootlegged batches of agave spirits to Mexico's overall liquor consumption likely dwarf all the legally certified tequila, *mezcal, raicilla,* and *bacanora* drunk within the republic. In a country where two out of every five bottles of distilled alcohol drunk remain outside the law, all the regulatory councils in Mexico cannot do much to move the needle.

Let us not forget one historic fact: For at least three centuries, mezcals have traveled more by mule, donkey, and horse than by train, boat, and plane. Their quality was "certified" by word-of-mouth ratings rather than anything on *Tripadvisor* or *Mezcal Reviews.* Many mezcals still travel humble, more wayward trails from their roots in the Sierra Madre Occidental and the Trans-Volcanic Belt in Mexico to the "outer world." Well into the last century, these mezcals were made in small batches for direct sales to favored customers, for they were bootlegged in mobile stills or *trens* hidden in canyons, rincons, and barrancas in the hinterlands of Mexico. That has been part of their mystique, just as it has been for moonshiners of corn likker in Appalachia.

These *mezcaleros* steadfastly refused to obtain permits nor pay any taxes or tariffs of any kind, though some paid extortion fees to unscrupulous law enforcers so that they would look the other way. In that manner, mezcal had good company, for at least two-fifths of all the distilled alcohol in the world remains made by clandestine bootleggers.

In Fausto Rasero's reflection, "Of Wisdom and Eternity" in *Mezcal: Arte Tradicional,* he links mezcal's legacy to a history of political and economic oppression as well as spiritual suppression: "Banned during colonial times for its connection with pagan festivals and its interference with brandy and wine imports from Spain, mezcal was only recently able to shed the stigma of its illicit, humble origins; for this reason, its distribution was limited during centuries to rural areas. Globalization has led to a greater awareness of traditional ways of life around the world, thus creating a paradoxical tendency toward homogenization. This has also meant that anything with a unique essence and deep cultural roots is assigned a surprisingly high value."

We must admit that as young men, we found romance in risking a petty crime. It fueled us with a certain sense of adventure as we chased down bootleggers and sought to sample the best of their mezcals. That quest excited us enough to risk run-ins with a local constable or sheriff to find some "homestyle hooch" wherever we went in Mexico.

Gary remembers a moment in the mid-1970s when he and his compadre Tomás rolled into the little riverside pueblo of Cucurpe, Sonora, just before dark, hoping to find some liquor that would warm their throats while camping in the cold that December. Locals directed them to the "dark end of the street" on the poor side of town, where humble families of Indigenous descent still occasionally sold *mezcal lechuguilla, bacanora,* and sotol. They desperately needed these sales as an extra source of income over and above what they brought home from their cutting of firewood, branding of cattle, harvesting of wild chiles and acorns, or leatherwork for making ropes, reins, and quirts.

After looking for where lantern lights were on in small adobe homes, Tomás and Gary went door to door asking for directions to a *mezcalero* that might have some extra spirits to sell. At last, an older woman directed them to the home of her brother, whom she claimed made the best *mezcal lechuguilla* in town, but also harbored a few bottles of infused *bacanora* and aged sotols from their kin who lived closer to the Chihuahua border.

When they arrived at the door of a run-down *jacal,* there was no electric light emanating from the house, only the occasional flash of firelight in a smoke-filled room from what must have been the hearth used for cooking. They knocked, but there was not an immediate answer.

They knocked again.

Through closed Dutch doors, a man's voice asked in Spanish what business they had. In Spanish as well, they responded that some shots of mezcal warming their throats would help them survive the cold overnight. The man, still unseen and suspicious of his visitors, asked plainly if they were police. They assured him that no, they were simply fans of *bacanora* with no solid ties to the local nighttime distilling scene.

"Well then," the hidden figure said, opening the top half of his Dutch door and extending a wrinkled arm, "give me your bottle."

Gary and Tomás gave him the only bottle they had in the car, an empty liter of Carta Blanca beer. The house was dark aside from the glow of the hearth, and as his wife held the bottle, the old man slowly filled it to the brim from a large demijohn that he tilted over it, consulting the beams from our flashlights to make sure he did not spill much. His wife then nestled a hand-carved cork into the mouth of the bottle, dried it off on her apron, and handed it out the Dutch doors. The man asked for a 100 pesos, or 50 plus the flashlight. It was a deal!

That was just the way *mezcal bacanora, lechuguilla,* and sotol had been sold for decades in northwest Mexico. These spirits had gone underground with the Ley Seca ("dry law") proclamation in Sonora on August 8, 1915, when then Governor Elías Calles was hell-bent on destroying his state's alcohol industry within five years' time.

The teetotaler failed to comprehend the resistance that the sweep of his ink pen would meet over the next eight decades. What Elías Calles engendered was not sobriety, but a resilient, clandestine mezcal industry of a kind that other states in Mexico came upon as well. When, on January 17, 1920, the United States enacted its own Prohibition law that lasted until 1933, the *mochomos* or "night ants" of Sonora carried tens of thousands of liters of Mexican moonshine across the border at night, igniting fame and lasting appreciation for *maestros mezcaleros* in the border states.

Simply put, mezcal first became something special in the Roaring Twenties because it was triply illegal—clandestinely made in Mexico without the payment of taxes or registration fees, smuggled across the border without payment of duties, and drunk in unregulated Southwestern speakeasies without any taxes going to the US government.

Since the Prohibition eras in Mexico and the United States, much has changed along the supply chain for distilled drinks and their feisty producers. The aging of mezcals before they reach consumers has become standardized and codified, but at a price.

Most of the mezcal that reaches US border towns to this day remains *mezcal joven*, that unaged, translucently clear distillate that often has a bright if not sharp, peppery bite to it. And most margaritas continue to be made with cheapo, adulterated *mixtos* of tequila.

But we now have far greater access to those *reposados* that we mentioned earlier, mezcals aged in oak barrels for at least two months, so that the tannins of the wood confer a caramelized hue to the liquor. Despite dozens of native oak species in Mexico, the preferred oaks for barrels continue to be the less porous, more aromatic *roble blanco* (white oak, *Quercus alba*) and *encino* (evergreen oak, *Quercus ilex*), both imported from the Mediterranean.

We have far more choices of *mezcal añejos*, aged for at least a year in volumes of 200 liters or less, in barrels usually imported from Portugal, Spain, Italy, or Greece. The *añejado* process deepens the distillate to a dark caramel color while smoothing and mellowing the fragrances and flavors.

Remember, however, that Mexican *mezcaleros* have been storing *añejados* in their own handmade barrels from other woods for decades, often to commemorate the death of a great *mezcalero* by taking out a few *pistos* to share among his friends on each death anniversary. A little of the distilled juice is often spilled and offered to a statue of a martyred *santo*, a photo of the deceased, or a drawing of the crucified Christ, just to remind everyone that few of us get out of this world alive!

By definition, regional specialties such as *bacanora* could also be termed *mezcal abocados*, in the sense that they have been often infused with natural colorings or aromas from fruits and herbs for months. Most Oaxacan *pechugas* could also be called a *mezcal abocado,* as could the *mezcal de gusano,* with its larval celebrity enlivening the rotund bottle.

But all mezcals, aged or otherwise, follow two rules. For one, mezcal must be bottled within the Denomination of Origin. No matter if it is unaged or spent fifteen years in a barrel, it must be bottled where it is produced. In contrast, *mixto* tequila can be shipped as concentrate to the United States in a tanker, then diluted and artificially colored and flavored at the US bottling facility.

Secondly, all mezcal is by regulation 100 percent agave. There are no *mixtos.* Unlike tequila, where producers can use only one kind of agave but are unrestricted in the types of other sugars they add, mezcal can be produced with as many types of agave as a producer wishes—but only agave, and only agave grown in the Denomination of Origin. This is to emphasize: 100 percent agave is not a distinguishing factor in mezcal. Alcohol content, however, is. If the alcohol strength is below 45 percent—even though 36 percent is legally allowable—ask the seller for an explanation of what is going on. The alcohol strength for a good mezcal is typically between 45 and 55 percent.

❋

The other issue that tarnished the industry after the "Mezcal Boom" in the 1990s was the bottling of mezcals—often blended from several cheap sources—labeled as *envasado en Mexico.* You can be assured of more authenticity and distinctive quality by sticking to the ones labeled as *envasado de origen,* for they are bottled at the distillery where the agaves were roasted, crushed, and fermented. Before it was curtailed, the off-site relabeling of tequilas for boutique, celebrity-endorsed niche markets became a folly in the early 2000s. Movie stars George Clooney and Dwayne Johnson, model and reality TV star Kendall Jenner, tech mogul Elon Musk, basketball's Michael Jordan, as well as musicians Justin Timberlake, Sammy Hagar, and Roger Clyne have tried their hand at custom-labeling versions of Mexican moonshine. They did so as if their pretty faces and fat wallets were sufficient to make the spirits taste more delicious.

With over 250 for-export brands of mezcal departing Mexico every month, it is hard to track where all of them go, and under what secondary labels. We do know that at least two-thirds of them are destined for the United States and Canada. But even in this era of scanning lot codes, the export of a Mexican mezcal or tequila to its neighbors to the north can be fraught with peril.

✳

Most aficionados of mezcal already know that producers, exporters, and border brokers are regulated by something called Mexican NOM-070-SCFI-1994 (NOM is the Norma Oficial Mexicana, which denotes that the bottle is authentic agave spirits produced in Mexico and tells you which producer it comes from). This law set the standards that put mezcal's Denomination of Origin in step with all the other international DOs, including the one earlier established for tequila. But DOs are but one of several place-based and culture-based protection strategies amidst a larger body of legal labeling protocols collectively called "geographic indications."

Keep in mind that the regulatory councils themselves do not set the rules and regulations for the production, certification, classification, labeling, and exportation that pertain to any of the agave distillates. These legal guidelines are established through Mexico's Secretaría de Economía (SE), via its Instituto Mexicano de la Propiedad Industrial (IMPI). The Mezcal Regulatory Council (CRM) was formed in 1997 to be the first (but not only) body to evaluate and certify whether producers and importers are following IMPI's laws and protocols.

In 2017, when IMPI authorized two other certifiers and verifiers to its list, the CRM suddenly realized it would no longer have a monopoly on executing those tasks. It had a temper tantrum unmatched by any other teenage brat on the planet. The CRM started to bully both its competitors and small-scale producers.

Then in June 2020, a moment of political infighting precipitated something altogether unprecedented: Mexico's federal SE sanctioned and fined the CRM 260,640 pesos for its "deceptive propaganda," which claimed it was the exclusive certifier of mezcals. The SE rejected CRM's public contention that "If an agave spirit does not have the hologram of CRM, it isn't mezcal!" The Secretary of Economy also noted that the CRM had "intentionally, repeatedly, and unjustifiably" refused to recognize or work with the other certifying bodies that were sanctioned in 2017.

Moreover, the SE fined the CRM another 695,040 pesos for intentionally deluding mezcal producers who wanted to use the labs of other legitimate certifying bodies in Michoacán, claiming that their certification could be accomplished only through exclusive contracts with CRM's own laboratory. The SE's ruling not only nullified the contracts forced upon small producers, but also freed the producers to certify with other entities. Federal government investigators released a damning picture of the CRM's bullying behavior toward producers and importers over the previous three years.

While the total fines imposed on the CRM were a pittance, amounting to less than 1 million pesos ($200,000), the SE sternly warned the CRM hierarchy that any more violations would result in the revocation of the CRM's authority to certify mezcal. The Mexican Secretary of Economy announced in no uncertain terms that the CRM would need to play fair or it would be kicked out of the game.

Mezcal blogger and tour guide Clayton Szczech rightly read the rebuke of CRM as a warning that the council should stay away from ever again hassling any producers, of either mezcal or noncertified agave distillates. The council members could no longer unfairly favor the mezcals they had been paid to certify over other agave distillates that had begun to compete in the national and global marketplaces. Besides, as Clayton quipped to us during a conversation we had at the Cascahuín distiller in El Arenal, Jalisco, "One hundred percent agave spirits not even sanctioned as mezcal by the CRM is really where it's at, and where it will be for some time."

The Secretary of Economy was clearly worried that such bullying campaigns could harm not only producers, but other certifying bodies, exporters, and consumers as well. And so, for the first time in history, the Mexican government had punished a regulatory council in Mexico for proliferating self-serving propaganda that was "imprecise, incorrect, exaggerated, partial, artificial, and biased." The CRM had disparaged if not dammed all attempts to dissuade importers and consumers abroad from sampling other agave distillates.

There has never been just one uniform mezcal to regulate, as there has been (to some extent) for tequila. As John McEvoy, the *Mezcal PhD* blogger has noted, a one-size-fits-all Denomination of Origin for the many mezcals "set [them] on a path to legitimization, but it was controversial from the start. One main point of consternation was that this law so closely mirrored the tequila NOM. Many producers and other stakeholders conceded from the beginning that this was a poor place to start."

The 2020 decision by the SE to chastise an abusive, overbearing regulatory council and its industrial enablers was a shot heard round the world. It was a game changer, because it publicly acknowledged that *mezcaleros* themselves needed to be granted the power to make decisions that shaped their own destiny.

Perhaps in another universe, we'd remember 2020 as the year of "Agave Spring." It initiated the liberation and sanctioned the re-diversification of all 100 percent agave distillates. They no longer had to fit one predetermined mold. It was centuries in the making—centuries of buried history, exploitation, and an extraordinarily resilient group of people and plants.

Chapter Eleven

᭞

LOST AGAVES AND THEIR FORGOTTEN SPIRITS

Something unprecedented began to appear with more frequency in both the press and social media around 2019 that has continued to this day: Journalists began to speak of the recent loss of certain beverage and food plants as "the extinction crisis that no one was talking about." Some called them "endangered foods and drinks," "gastronomic gaps," "lost feasts," or even "culinary extinctions." New lists proliferated of now-extinct species that had once broken the hunger or slaked the thirst of thousands if not millions of human consumers. The big question that these alerts prompted was whether any of these lost delicacies could be resurrected.

The answer is yes, but rather oddly, none of the lists mentioned any agaves among the current efforts toward "resurrecting" forgotten beverages and foods. In fact, some are being revived as we speak, but you will shortly see why it was easy to miss these shining stories of success among the haystack of extinctions that now occur every day.

In 1791, a Guatemalan-born natural historian named Lieutenant Antonio Pineda y Ramirez landed in western Mexico to join a royal expedition commissioned by the Count of Revilla Gigedo, Viceroy of New Spain. He went from the port of Acapulco overland to Mexico City to meet other members of the Malaspina Expedition, where they began several months of travel throughout Mexico, canvassing lands in the present states of Guanajuato, Guerrero, Hidalgo, México, Querétaro, and Puebla. Over the eight to ten months that he explored the Mesoamerican landscape for specimens of plants and wildlife, Don Antonio de Pineda took it upon himself to record all the fermented and distilled beverages of the Indigenous inhabitants of Mexico that he could find and sample at that time.

What stuns any reader of Pineda's report on "recipes" for seventy-seven different beverages is the amazing variety of drinks that employed agaves as their dominant ingredient. The lieutenant described more than thirty distinctive probiotic beverages that used aguamiel, pulque, or the roasted stalks, leaves, and hearts of agaves, fermented together with fruits and spices.

Pineda also described nine distilled beverages that utilized agaves, as well as three runs of *mezcals* that were separated and used in different manners, or combined in balanced distillates. Some distillates on his list sound very much like the *comitecos* of Chiapas that include aguamiel run through a still to obtain an aguardiente.

These ancient distillates of agaves have a range of alluring names: *bingarrote* (still produced in Guanajuato), *excomunión* (still produced in Michoacán), *mezcal de pulque*, anise-infused *mistela por alambique, aguardiente de pulque, vino mezcal de Guadalajara,* and *vino resacado*. Some were infused with herbs and fruit, others combined with the juices of sugarcane or *coco*, but all sound delectable.

Most of what Pineda learned of these recipes for employing agaves in various fermented and distilled beverages probably never even made it into print. He died while on another expedition within fourteen months of initiating his survey of the spirits of Mexico. And yet, he was so beloved by his comrades on the Malaspina Expedition that they made sure that two of his reports were published posthumously, and that a large memorial was erected in his honor in the Botanical Gardens of Malate, in Manila, then the biggest port town in the Philippines. It included in the inscription: "In three years of arduous journeys, he traveled to the ends of the world, exploring the bowels of the earth, the depths of the sea and the highest peaks. . . . The early death of this noble man is mourned by his countrymen, the flora, fauna, and friends who have erected the monument."

What we gain from Pineda's work is both a sense of wonder regarding the diversity and antiquity of agave distillates in Mesoamerica, and a sadness that some of those may now be lost from sight and taste forever. Keep in mind that Pineda was writing three centuries after Columbus and two and half centuries after the Conquista. He lived and lusted after liquors decades after Francisco Hernández did the same. Hernández, the other remarkable natural historian of Indigenous foods and drinks, was among the first to describe how the Chichimecas baked the leaves (*pencas*) and "trunks" (*mezontles*) of agaves in underground ovens called *barbacoas*, and then fermented and distilled their juices.

And so, there is abundant evidence from the so-called protohistoric period that agave distillation—not just fermentation—was already regarded as a worthy tradition. But with the depopulation and acculturation of Mesoamerica during the first century after Cortés, many Mesoamerican gastronomic practices—like eating popped amaranth grains mixed with fresh blood—were suddenly declared to be "heathen" and were aggressively disrupted or altogether banned.

It is impossible for us to know how many distinctive drinks of fermented and distilled agave juices fell out of fashion and into oblivion. Only through the all-too-brief glimpses we are offered by Hernández and Pineda can we even envision a world where innovation and tradition combined to foment dozens of regionally celebrated agave drinks.

As we began to compile a list of all the forgotten, lost, or rare agave distillates of Mexico recorded since the time of Antonio de Pineda, the list grew longer and stranger. It includes *bacanora* beyond Sonora and in adjacent Chihuahua, *bingarrote* of Guanjuato, *canoas* of coastal Jalisco, *chihuahualco* of Michoacán, *comiteco* of Chiapas, *excommun* of Michoacán, *huitzila* of Zacatecas, *jaiboli* of Sonora and adjacent Chihuahua, *jarcia* of Tamaulipas, *lechuguilla* of Sonora and Chihuahua, *minero* of Oaxaca, *petaquillas* of Guerrero, *quitupan* of Jalisco, *raicilla* (*de la costa* and *de sierra*) of Jalisco and Nayarit, *sikua* of Michoacán, *tasequi* and *tuata* of southern Sonora, *tepe/torrecillas* of Durango, *tlahuelompa* of Hidalgo, *turicato* of Michoacán, *tuxca* of Colima and adjacent Jalisco, *yocogihua* of Sonora and adjacent Sinaloa, *zihuaquio* of Guerrero, and *zotolero* of Puebla.

Even at face value, this list shocks us into reconsidering the tremendous diversity historically found in agave distillates. We long to cultivate, harvest, ferment, distill, or taste them. For the most part, these traditions have not been circumscribed within—nor made obsolete by—the operating instructions guiding the tequila and mezcal regulatory councils. Of course, many of these distillates suffered some setbacks well before the founding of these regulatory councils. And yet, the remaining fragments of certain traditions were roundly disparaged in some quarters, if not altogether banned by regulators.

While every one of these historic agave distillates has a unique story of its rise, demise, and potential rejuvenation, there are some themes shared by most of their narratives.

First, those who controlled the importation of European or North American spirits to Mexico wanted to eliminate Indigenous competitors and their homegrown "hooch" to control the New World markets for their relatives and business relations back in Europe.

Second, this suppression drove production underground within their localities of origin, such that Mexican distillers rarely sought out external praise or markets overseas. In a sense, their agave distillates became *vino mezcals de la casa* or *de la cantina* that stayed below the radar of most authorities.

To our surprise, such homemade agave spirits remain among the most widely drunk, clandestinely distilled spirits in the Republic of Mexico. As we noted earlier, liquors made, adulterated, or traded by bootleggers collectively comprise 43 percent of all the spirits consumed behind the Tortilla Curtain to this day.

Third, several Prohibition-like proclamations made by provincial viceroys, state governors, and republic presidents in Mexico were met with mixed success over the centuries. To be sure, the Ley Seca proclamations perpetrated by Elías Calles were brutal enough to drive hundreds of small-scale distilleries out of business, even as others went underground. We can only imagine that considerable traditional Mexican knowledge about growing, fermenting, and distilling agaves that was lost in the wake of those

Dry Times. To the extent that it concerns us at all, we see through the glass darkly to the agave spirits once held within it.

More recently, the Mezcal Regulatory Council (CRM) has sporadically tried again to render some of these distillates illegal, or at least deprive them from using the word *mezcal* anywhere on their labels. But at other moments, the CRM has endeavored to lure producers to work under its umbrella rather than letting them apply for their own Denomination of Origin. We might diagnose this contradictory behavior as bureaucratic schizophrenia, as regulators vainly attempt to retain control of marketing trends when the trend has already left the station.

As one example, when Indigenous Purépecha distillers of *sikua* tried to obtain their own DO, the CRM expanded its legal range to include their municipalities in Michoacán and invited *sikua* producers to use the existing DO for mezcals rather than start their own *consejo*, in order to capture more testing and verification fees from these producers.

Ironically, the high cost of entry to receive legal status of *sikua* agave distillates under the CRM virtually stifled the production of spirits from that distinctive tradition of the Pátzcuaro region. After 2016, the Purépecha distillers lost much of their previously "free" publicity and failed to gain a broader market for their products. *Sikua* appears to be more marginalized today than it was in 2010, as the CRM worked to recruit hundreds of Purépecha producers of agave spirits in the three municipalities that stretch along the shores of Lake Pátzcuaro.

To get a sense of what might be lost if such trends continue—and what might still be revived among rare and nearly forgotten agave distillates—let's delve into what happened with one of the lesser-known spirits of Sonora. This mezcal was made at the Yocogihua distillery near the Sonora-Sinaloa border from 1888 to 1965. The distillery was located between the historic mining town of Alamos, Sonora, and the Mayo Indian town of Masiaca, at the place along an arroyo called Yocogihua—"where the jaguar eats."

According to Alamos historian Juan Carlos Holguín Balderrama, the

Aldama brothers built a distillery on their ranch in 1888. They began to use wild agaves and two domesticated landraces of *Agave rhodacantha,* a towering mezcal that is grown for making spirits much farther south on the west coast of Mexico. This subtropical species has always been rather rare in Sonora. In fact, it is known from two pressed herbarium collections near Alamos and just one other locality over 100 kilometers to the north. However, it flourished in the Aldama brothers' plantations along the Río Masiaca, where they called their two special agave varieties El Chino and San Antoneña.

Some records suggest that *Agave rhodacantha* was initially brought northward from Jalisco in large numbers in a well-organized effort to ramp up agave cultivation in northern Sinaloa and southern Sonora. The goal was to build a mezcal industry that would export liquor by rail to the US.

Around that time, the Sonora Railway—a branch line of Edward Harriman's Southern Pacific Railway—had finished laying down its tracks between nearby Guaymas, Sonora, and Nogales, Arizona, in 1882. A sizable investment by "robber baron" Harriman in constructing five interconnecting railroad lines opened the American West to the gastronomic treasures and pleasures of the subtropics of Mexico. Within a quarter century, Harriman had made enough profits hauling Mexico's treasures into the US that he acquired the little overland stagecoach mail line in the border states that he turned into the Wells Fargo banking empire.

And so, to vanquish the thirst of railroad workers and passengers alike, the Yocogihua distillery became one of sixty-six commercial *mezcalerías* or *vinatas* in Sonora in the 1890s that vied to place its demijohns on railroad cars going up toward the US border. The volumes of tequila and mezcal exported to the US grew exponentially. Within a decade, the Yocogihua distillery employed about a dozen cultivators and harvesters and five other laborers, producing nearly 19,000 liters of various agave spirits per year.

If observations and oral histories heard by Howard Gentry in the Alamos region are correct, this and other distilleries nearby rapidly depleted three wild species of agaves within a matter of just a few years. However, the cultivation of the two *Agave rhodacantha* varieties on the plantation flourished. It is likely that the Aldamas employed Mayo and Guarijio Indian workers who had a long tradition of utilizing more than ten different species and varieties of agaves for drink.

Around that time, Plutarco Elías Calles made a fateful declaration in August 1915 from the capitol in Sonora. Some historians claim he did so because a drunken father or uncle had beaten him when he was a child. But others suggest that his grandfather was a teetotaling Syrian (and perhaps of Muslim origin), having been recruited from the Middle East by the Texas Camel Corps in the 1850s before he ventured south into Sonora to marry an Indigenous woman in southern Sonora. In any case, Elías Calles made the agronomic production, distillation, sales, or possession of mezcals crimes punishable with a five-year term of hard labor. Most violators were jailed, but some repeat violators were reputedly murdered as Calles set out to eradicate all forms of moonshine made in his state. Unable to maintain their ranch and its agave plantations without their former income from mezcal sales, the Aldama family descendants sold it off in 1927.

Then, in 1934, the ranch was sold to an enterprising Yugoslavian immigrant who had recently arrived in the region. Calles soon left Sonora to become commerce secretary, then interior secretary, then president of the republic, and finally, a brutal behind-the-scenes power broker who kept three other presidents under his thumb. It was not until 1934 that he was forced to take exile in California.

About that time, the new owner of the ranch, David Andrés Sugich Rafaelovich, was audacious enough to open his first liquor store not far from Calles's old backyard, in Navojoa, Sonora. With the atheist teetotaler permanently out of the picture, David Sugich seized the opportunity to capture the distribution of Modelo beer across all of Sonora, to open 134 cantinas statewide, and to find semilegal means to jump-start production of mezcal in the old distillery.

Sugich quickly renovated the old agave fields once tended by Mayo cultivators and harvesters. Historian Juan Carlos Holguín helps us imagine the scene at that time. The ambitious immigrant built out a mechanical agave shredder, twenty-three fermentation vats of 800-gallon (3,100-liter) capacity, three boilers, alembic stills, a well, and two storage reservoirs for fresh water.

The agaves used in the second incarnation of Yocogihua mezcal were largely grown on the ranch itself, where the San Antoneña and El Chino varieties had proven to be well adapted to the arid subtropical landscape. Each hectare harbored 2,500 agaves cared for by eight to ten men, who weeded, pruned, and harvested the crop for mezcal, while propagating new offshoots in nurseries. Their mezcal production began once the mature agaves were cut, with the harvest carried out by seasoned cultivators and harvesters backed by a few day laborers. Once the *piñas* were trimmed, they were loaded onto mule-drawn *carretas* and taken to the *vinata*, where they were split in half. They roasted in two masonry ovens for seventy-two hours. Each brick oven had a capacity of 10 tons of trimmed agave.

Historian Holguín reconstructed their process: "Once cooked, the roasted heads were passed through a shredder and then crushed in a *tahona*, where the extracted juice was siphoned into vats where they were left to ferment for five to seven days. Once fermented, they passed this must into the alembic still, which was initially fueled with firewood and later with a carbon-rich tar. The still had a capacity of 1,000 liters per batch. It took twenty-four hours to double-distill and balance the mezcal, before it was sent in barrels by wagon for bottling in Navojoa."

There is little doubt that hush money payments ended up in politicians' pockets, for Sugich's ranch was soon employing twenty workers as "vaqueros" who probably did raise a few cows, but mostly assured that 200 liters of Yocogihua mezcal were produced each day. After bottling in Navojoa, the mezcal traveled out on trains or trucks through Sugich's distribution network. It had grown to serve hundreds of bars and *depositos* in Sonora, Sinaloa, and Baja California.

When David Sugich left his mezcal-drenched world in March 1965, the distillery continued for a few more years but no longer had his old distribution network to bars and liquor stores beyond Navajoa, Obregon, and Alamos. The ranch's mezcal production went into steep decline, and the business fell into debt.

According to historian Holguín, federal authorities expropriated the 6,490 hectares of Yocogihua Ranch in 1975, perhaps because the *mordida* payments had stopped to the politicians who had allowed decades of agave spirits production despite the Ley Seca. With the plantation in ruin and the wild agave populations in the surrounding Alamos region badly depleted, the distillery permanently closed its doors in 1985.

That happened to be the last year that Gary was ever able to find a bottle of Yocogihua in the town of Navojoa. The brand could still be purchased at one of Sugich's old liquor stores that had the Yocogihua label emblazoned as a mural on the wall. Within a decade's time, all the remaining stock of the Yocogihua distillate in warehouses was depleted, the ranch was left in disarray, and this unique agave tradition fell into oblivion. It appeared to have become an extinct drink, a forgotten spirit.

And yet, the story of Yocogihua was not over.

Over decades of visiting the Alamos region of the Sierra Madre Occidental, Gary has stumbled upon several distinctive, desert-adapted varieties of *Agave rhodacantha* in Mayo Indian rancherias and pueblos not far from the ruin of the old Yocogihua distillery. Two of the only three collections of this species recorded from Sonora are his herbarium sheets made from sites within an hour's drive to the old distillery.

Could these be survivors taken from Rancho Yocogihua, where the plantations around the distillery once stood? Could they be remnants of old stock rescued by the last Mayo or Guarijio Indian families working at the Aldama-Sugich family ranch on the Arroyo Masiaca? Were they informally "repatriated" to the Indigenous cultivators and harvesters as the distillery fell into ruins? Could they be the living bridge to the San Antoneña and El Chino folk varieties that the Aldama family once grew for mezcal in southern Sonora?

Would it be worth gifting these folk varieties or landraces of rare agave to some *maestro bacanorero*—like Ramón Miranda Urrea and his son Eduardo Miranda—who make the best agave spirits in Sonora at their El Real de Alamos distillery? That *vinata* lies less than an hour's drive from where the Yocogihua distillery once functioned. If Ramón and Eduardo chose to grow them for distillation and history's sake, Sonora could regain its historic status of having several agave distillates—not just *bacanora*—in the marketplace.

The work to recover these lost agave varieties and return them to distillers has now begun. Expeditions to the Yocogihua region below the Sierra Madre of southern Sonora are in the planning stages. We cannot yet guarantee that you will taste a resurrected mezcal soon, but are sure it will happen.

⁂

Possibilities like this are already in process in other mezcal regions of Mexico. Ana Valenzuela spent more than three decades talking to elderly *jimadores* to help her track down the seven "heirloom" agave varieties that were once gown alongside the *tequilana azul* cultivar in the Altos and Valles regions of Jalisco. She scoured hedgerows, scattered patches in front of distilleries, and remnants in abandoned fields until she recovered the surviving genetic stock of the landraces of folk varieties known as *bermejo, listado, moraleño, pata de mula, sahuayo, sigüín,* and *zopilote.*

But Ana went well beyond treating them as museum pieces. She carefully evaluated the unique traits of each of these heirloom agaves and proposed a future place for each of them in her classic manifesto, *A New Agenda for the Blue Agave.* They can now be found in any one of several conservation gardens, where a total of fifteen historic mezcal varieties formerly thought to be lost are being tended by scientists and farmers. One of them is just outside of Zapotitlán, Jalisco.

The same kind of process has been initiated by *mezcaleros* themselves on other parcels of land that are also in sight of the Volcan de Colima. That is where the ancient *tuxca* tradition was historically described around the pueblo of Tuxcacuesco near the present-day Jalisco-Colima border. To this day, old-timers still sometimes call their spirits *tuxca* when they sell it on Sundays with a *ponche* in the market and roadside stands of Tuxcacuesco. The great *maestros mezcaleros* of that region—Macario Partida, Tomás Virgen, and Santos Juárez—still associate some of the older cultivated varieties with that *tuxca* tradition, informally calling the plants themselves or their resulting spirits *tuxcas.* Men like Don Macario and Miguel Partida go on scouting expeditions in the barrancas between the Nevado de Colima and the Sierra de Manantlán to seek out feral varieties that they can bring back to their fields for evaluation.

This is a form of plant conservation that biologists have dubbed *resurrec-*

tion ecology. Or in this case, maybe we should call it *resurrection ethnobiology.* It is the discovery and revival of forgotten sources of plant genes and flavors. The goal is not at all "retro"—like promoting Coke Classic—but instead, to employ them for novel uses in a new era. These agaves are like the proverbial phoenix birds, with new life rising out of the ashes strewn out in the fields around the roasting pits of a traditional distillery.

⁂

The families Partida, Juárez, and Virgen employ many of the same distillation traditions, flavor sensibilities, and traditional knowledge of plant care once used by the historic *tuxca mezcaleros* of their grandparents' era. But today, they must call their products "100 percent distillates of agave," for they can no longer legally market them as mezcal or *vino mezcal de Tuxca* because their lands fall outside the permitted range of production delimited by the Denomination of Origin for Mezcal.

But what if they simply used collective trademarks, geographic indicators, or appellations of origin to name and promote the historic legacy of the *tuxca* tradition? The traditional processes and the distinctive agave landraces would still be there. Diversity would increase, rather than continuing to decline inexorably.

Indeed, this may be what is happening with the *minero* tradition in Oaxaca, the *bingarrote* tradition in Guanajuato, the *tepe* tradition in Durango, and the *lechuguilla* tradition in far western Chihuahua. The old "heirloom" varieties simply need to be put back in the hands of the best *maestros mezcaleros* working within these local traditions. These producers can incorporate as a nonprofit civil association that can set their own criteria for maintaining the quality of their 100 percent agave distillates without having to consult with the Consejo Regulador del Mezcal. Applying for a *marca colectiva* would be enough to propel them on their way.

As Patricia Colunga-GarcíaMarín continues to remind the mezcal aficionados of agave spirits who have ears and minds open enough to hear her, "The futures of agave spirits will emerge from our ancestral traditions, not from industrialized processes and conventions." This is where grassroots action will meet the heart of the agave spirit world.

Chapter Twelve

THE WILD DANCE BETWEEN
AGAVES AND BATS

While we love celebrating Howard Gentry's notion that agaves and humankind have entered a special kind of symbiotic relationship with one another, we do not wish to ignore the other symbiotic relationships that keep agaves fruitful and tasteful.

One such symbiosis works on the night shift. It may happen when few of us can observe it, but that's true for most acts of reproduction and regeneration, don't you think? In this case, we are referring to the coevolutionary dance between bat-friendly agaves and nectar-feeding bats. To glimpse the underpinnings of this ancient set of ecological interactions, we traveled to see a friend down in the "soul of Mexico," Michoacán, in what was once known as the Tarascan Empire of the Indigenous Purépecha.

Our first stop on Rancho El Limón was Emilio Vieyra Rangel's nursery of century plants. Not to his home, his *vinata* distillery, nor even his restored forests of maturing plants. He took us to his *vivero*.

With the pride of a father, he walked us down the paths among the thousands of seedlings of *Agave cupreata*. He tipped back his cowboy hat, knelt on the knees of his blue jeans, and pointed out to us the amazing variability in array of sizes, shapes, and colors in his hand-seeded century plants.

"This *Agave cupreata* goes by many common names or aliases," he explained, as he stroked his goatee and grinned. "Some call it *cimarrón*, the wild one. Others use the old Aztec term in Nahuatl, *papalometl*—the butterfly agave, or *mezcal de mariposa* . . ."

Emilio gestured, turning his interlocked hands into the undulating wings of some butterfly, as if letting them sail like the now-endangered monarchs butterflies that come to roost for the winter in declining numbers in the nearby mountain ranges of Michoacán. He noted that still others call this agave *tuchi*, or *yaabendisi*—a Mixtec term.

Agave cupreata is one of the two succulent species Emilio grows in reforested areas for producing his family's artisanal mezcal, while another wild agave grows on its own, out of reach, clinging to the cliffs above his fields.

Not all agaves in Emilio's restored woodlands look or even taste the same. We could see that *cupreata* itself was irrepressibly heterogeneous. This is the kind of rainbow of colors and characteristic diversity that we can hardly see at all in the extensive plantations of blue tequila agaves 185 miles (300 kilometers) north of Emilio's home.

The persistence of such genetic diversity at Rancho El Limón owes much to the fact that Emilio's agaves are seedlings of *open-pollinated* plants— ones that are freely accessible to lesser long-nosed bats and other pollinators that carry pollen from one agave plant to fertilize flowers of another. They come to be nourished by the nutritious nectar in their flowers, even as they deposit viable pollen from other wild agaves that grow in the Río Balsas watershed.

To foster this ancient bat-agave mutualism, Emilio is now cultivating another kind of symbiosis between his own family and the agave family. He now leaves 3 to 5 percent of the *cupreata* agaves in his pine-oak woodlands untouched, reserving most of them to nourish imperiled bats rather than simply harvesting all of them for his roasting pit.

In doing so, Emilio and his family have become the pioneers of the bat-friendly mezcal and tequila movement, which began around 2015. This initiative is now being promoted in several states by David's Siembra Suro Foundation and its Tequila Interchange Project, and is being monitored on the ground by Dr. Rodrigo Medellín and his students at the National University (UNAM). It has also been endorsed and promoted by the many other wildlife conservationists that Rodrigo has trained in his role as the "Bat Man" of Mexico.

It is on Emilio's ranch that the project bore its first fruit. Emilio's family brand—Mezcal Don Mateo de la Sierra—not only includes the first set of artisanal mezcals to be certified in the state of Michoacán; they are also the first bat-friendly agave spirits to be verified in Mexico.

Emilio's conservation efforts are not some green-washed marketing ploy devised to simply attract nature-loving consumers to his products. The restoration initiative springs from his family's deep commitment to harboring more agaves and bats on the land than were there when he inherited Rancho El Limón after the untimely death of his father.

It is also an act of resistance against the dominant trends in both the tequila and mezcal industries over the last quarter century. Most tequila- and mezcal-producing agaves used for spirits are propagated vegetatively, rather than from seed. On Rancho El Limón, most of the young plantlets out in the restored forests were germinated from seeds derived from bat-pollinated flowers, not vegetatively cloned from a single mother plant.

Sadly, the clonal propagules that dominate the production of *tequilana azul* and *espadín* mezcal are seldom if ever allowed to mature and even flower so that they can be visited by pollinating bats. Today, because most blue tequila agaves are micro-propagated through tissue culture biotechnologies in petri dishes lined up on shelves in laboratories we have described earlier, we sense that sexual reproduction by "consenting" tequila plants is now an all-too-rare event. Most *tequilana azul* propagules come from a few closely related homogeneous clones that now cover hundreds of thousands of acres in four states to the north of Rancho El Limón.

Tequila production may have become one of the most genetically uniform beverage agri-businesses on Planet Desert, but mezcal production is struggling to take another, more benign route into the future. Emilio's management of habitats for wild agaves offers a necessary counterpoint and a whole other way of tasting and seeing the world that both producers and drinkers of tequila should deeply consider.

⁜

On our second day with *la familia Vieyra* on Rancho El Limón, Emilio took us aside to explain how the use of these genetically variable, semi-domesticated mezcal agaves began at the insistence of his father, Emilio Senior. Years ago, the father and son began to transplant out thousands of *Agave cupreata* and *Agave inaequidens* on their forested ranchlands. They also established an 86-acre (35-hectare) Voluntary Conservation Reserve above their agave agroforestry restoration plots that is protected from both livestock raising and commercial harvesting of plants of any kind.

If we were to unintentionally fly over or stumble upon the Vieyra lands we visited with Emilio, we would be hard-pressed to even recognize it as a place of agricultural or agroforestry production. The agaves are transplanted into the open woodlands, tucked in between pines, oaks, feather trees, and arborescent prickly pears, rather than being placed in neat rows spread out across plowed ground. Orchids, pineapple-like bromeliads, and mushrooms decorate the trees like so many holiday ornaments.

The wildness of the land has been restored, and the agave harvest is but a by-product of that generous human gesture of habitat restoration. No tillage, no pesticides, nor any herbicides are being used. Emilio's ranchlands that are dedicated to agave production look more like natural, wild habitat than manicured fields or irrigated pasturelands. What's more, wildlife—including bats—favorably respond to its naturalistic character.

Many who came along in our traveling band of mezcal aficionados believe that the mezcals distilled from the agaves grown in Emilio's restored woodlands seem to taste sweeter and wilder than the mezcals coming from other century plants growing in Michoacán.

⁜

Rancho El Limón's woodlands are chock-full of wild plants that are close relatives—if not direct ancestors—of the many food and beverage crops that have nourished Mexican families for nearly ten millennia. That is because

Lesser long-nosed bat pollinating agave blossoms

their watershed—that of the Río Balsas, with its deep, water-carved barrancas—is now considered by many scientists to have been the very cradle of
Mesoamerican agriculture. It is also a global hot spot for wild crop relatives,
the unruly kin of the cultivated plants that nourish us day in and day out. At
least 330 species of wild crop relatives in thirty genera of plants occur in or

near in the Balsas watershed. Because many of them have adaptations that might help our food crops endure climate change, they have become global priorities for conservation.

The canyon bottoms here truly cradle the wild corn-like grass called *teosintle*, as well as wild bean and gourd vines. Those persisting in the Río Balsas are suspected to be the direct progenitors of the corn-bean-squash staple crop triad known as the Three Sisters. Along with agaves, chiles, and prickly pear cactus, these botanical sisters form the origins of the Mesoamerican Diet that is now celebrated as a Biocultural Patrimony of Global Significance by UNESCO.

Thanks to the vision shared by Emilio Senior and his son, the copper-tipped, ecologically wondrous *Agave cupreata* and many other wild relatives are recovering. This agave is now abundant enough to offer its beautiful juices to those who helped enhance its populations. In turn, its abundance has helped a once-endangered pollinator species recover—the lesser long-nosed bat. It is a migratory bat that seasonally frequents the caves in nearby barrancas and the nocturnal blossoms of actively blooming agaves on cliffs.

Although the flowers are open when hummingbirds visit them on long summer days, the agave pollen is viable for fertilizing other flowers only during the night. Even pollen that the hummingbirds or bees move from plant to plant during the day may not be viable, further exacerbating dependence on nighttime pollinators. In welcome news, the famed "Bat Man" of Mexico has performed nocturnal field studies to confirm that the lesser long-nosed bat community on Rancho El Limón is sufficient to maintain the recovery. The bats do the lion's share of pollination for the agaves there, and Emilio attributes their abundance to the unrelenting efforts at stewardship taught to him by his father.

The bat-agave symbiosis is augmented by the human-agave symbiosis as a means for keeping wild agaves thriving and flavorful mezcal in our *jicarita* cups and shot glasses. But what is the nature of that more ancient mutualistic, coevolved relationship between bats and century plants that has been sustained for so many centuries?

For starters, look at the shape and size of a female lesser long-nosed bat, and the way she flies from Michoacán, Colima, and Jalisco up to Arizona and Sonora on her 1,500-mile (2,400-kilometer) migratory cycle each year. She has a snout that is far more elongated than other bats in her range, for she must poke her head into the deep tubes of agave and cactus flowers while hovering before them. She also has a rather long, brush-tipped tongue for sweeping up nectar and pollen from the pale night-blooming agave blossoms. In fact, she can live entirely on a diet of nectar and pollen year-round, if need be, something that other bat visitors to agave cannot necessarily do. They must rely on insects as well.

For their part, agave blossoms release copious amounts of dilute, sugar-rich nectar and protein-rich pollen at night to attract bats into moving their pollen from flower to flower. Flowers that coevolved with bat pollinators often have only a third the sucrose but more hexose than do flowers exclusively pollinated by bees. Hexose is associated with a musky scent easily discernible by bats at night, and can lure them in to a cluster of blossoms from considerable distances. And of course, there may be parallels between the shape, length, and breadth of the bat's snout and the tubular blossom of the agave.

Even the bats' groovy, bumpy tongues are highly adapted to collecting pollen and nectar, perfectly formed to glean these resources for the entire duration of a visit to a blossom. It is a special mechanism for fluid uptake not seen in any other mammal in the world.

Nectar-feeding bats are strong fliers, but also have the capacity to delicately hover and slowly move from flower to flower as each blossom leaks out nectar into its tube. Lesser long-nosed bats have been seen visiting the same cluster of agaves flowers two to three hundred times in a single night. They can easily forage out from their cave roosts 10 to 20 miles (16 to 32 kilometers) on any single night of their migration but might need the nectar from as many as five hundred flowers to fuel their entire two-way migration.

The nectar-feeding bats that visit agave flowers from mid-summer through fall have a strong sense of smell and vision good enough to spot a few pale maguey blossoms against a night sky. Agaves make their flowers easier to spot in desert and tropical vegetation by hoisting them up on towering flower stalks that position their blossoms as much as 48 feet (15 meters) above the ground. It takes an awful lot of energy to build the towering stalks that raise their flowers higher than the other plants in the arboreal deserts, oak woodlands, pine forests, and tropical jungles of North America. Perhaps the reason that they have erected the tallest flower stalks in the world is that they serve as signposts for stopovers along the bats' migration down their nectar corridor.

Overall, most wild agaves need to get good seed set so their populations can survive, so they have not only welcomed visitations by bats, but reshaped their architecture and floral rewards to that end.

There have been many challenges in keeping the symbiotic dance among agaves and bats alive so that it could make it into the new millennium. Some of the efforts of Emilio and others clearly seem to be working, for they have tangible conservation value. We now stand in a historic moment in which once-endangered bat populations appear to be recovering, and agaves are being replanted by the millions in both Mexico and the US Southwest.

The Tequila Interchange Project, UNAM's Institute of Ecology, Agave Power, Mexico's Sembrando Vida initiative, Heifer International, the Wild Agave Foundation, Borderlands Restoration Network, Colectivo Sonora Silvestre, and Bat Conservation International are among the many players that Rodrigo and David have inspired. Despite the initial successes of such initiatives, we would do well not to forget just how vulnerable both agave populations and bat roosts remain.

As early as a century ago, there were already warning signs that alerted any *mezcalero* who was paying attention that bats were getting scarce and that the overharvesting of agaves for mezcal had something to do with it.

A little parable from the era of the Mexican Revolution can help jump-start our understanding of the attractive powers and the impending perils of mezcals distilled from wild agaves. Before that oppressive Ley Seca was established, several different kinds of mezcal were being cultivated on small farms, often adjacent to the ones harvested from the wild.

As we have noted elsewhere, Sonorans sheltered *bacanora* in the frost-free subtropical thornscrub of the Sierra Madre, two cultivated *Agave rhodacantha* varieties (for *yocogihua* mezcal) in the transition between subtropical scrub and the tropical short tree forests, *lechuguilla* in the *mezquite* grasslands at the desert's edge, and a variety of other, more localized mezcals like *tauta, vicho,* and *jaiboli* in smaller patches.

Just before Elías Calles closed down all the plantations of agaves being grown for distilleries in Sonora, the harvesting of wild agaves and the cultivation of domesticated ones rendered 400,000 to 850,000 liters of spirits to soothe the parched mouths of dwellers in the Sonoran Desert.

By calculating the number of mature agaves needed to produce 850,000 liters of distilled alcohol without adding any cane sugar to the fermentation vats, we have estimated that at least a million slow-growing agaves were being harvested in Sonora annually up through 1910. Just as the Mexican Revolution was ramping up, there was already so much demand for alcohol from both sides of the civil war that *maestros mezcaleros* in Sonora found that the wild agave populations within their reach were suddenly in short supply.

There was seething social unrest in Sonora at that time. It reached a crescendo following mass killings of miners during a strike in the mountain town of Cananea, Sonora, in 1906. That tragic event is one of a handful that is said to have ushered in the Mexican Revolution.

When all-out warfare exploded in Sonora around 1910, the number of legal *vinatas* or distilleries suddenly dropped by a fifth. So mezcal-making went underground, and a thousand microdistilleries blossomed like desert wildflowers in the years just after the revolution.

No longer able to cultivate domesticated agaves, the *mezcaleros* sought out secluded canyons away from roads to do their distilling. Their favorite places for hiding their stills were canyon bottoms below springs that were in the midst of large wild populations of two or more species of agaves. There, with a mule and machete, they could work clandestinely in making bootleg agave "hooch."

That is when the bootleggers became known in code language as *mocho-*

mos, a colorful Sonoran metaphor that likened them to the leaf-cutter ants that harvest leaves from desert plants by day to ferment and cultivate them in underground gardens with the help of fungi at night. Like the nocturnal ants that were among the animal world's first "farmers," the distillers could ferment and then concentrate the potency of their fermented stores of plant carbohydrates to consume over the rest of the year.

Although the wild agaves in Sonora were often smaller than the ones cultivated before the revolution, they were divinely flavorful and could be obtained without the help of the sizable workforce needed to run an agave plantation.

And so, for the next eighty years, nearly every Sonoran boy was recruited by his father or uncles to labor for a while in the making of bottled mezcal. It became a rite of passage or badge of honor for every *Sonorense* who lived without refrigeration for cold beer to help his father and uncles produce the mezcal reserves that would be needed over the coming year.

Because of the added cultural value that a bottle of bootleg *bacano*ra had over tequila or Oaxacan mezcals, the number of stills in Sonora likely increased during the years of Prohibition.

Since the stills themselves were mobile, bootleggers could pick up shop and hide somewhere else if a sheriff or constable asked too much as a protection or extortion fee. But the mobile stills allowed some agave plant populations to recover from harvesting pressure before a return visit occurred. In addition, most of the harvesters in those days were trained by older *mezcaleros* to "prune" rather than destructively harvest entire plants connected by cord-like rhizomes underground, so that agave populations did not necessarily decline.

All these traditions began to shift thirty years ago. By the early 1990s, as the global interest in mezcals began to rise exponentially, the Sonorans abolished the constraints of Prohibition. In fact, the state government began to promote *bacanora* as a legal product, following in the path taken by producers of *raicilla* in Jalisco and sotol in Chihuahua.

It was during that time that ecologists first noticed serious signs of overharvesting in Sonora. The new initiatives for legalized markets incidentally precipitated the removal of as many as two million plants per year, either for transplanting to fields around renovated distilleries, or for roasting on a scale not seen for decades. Distilleries were no longer mobile stills, and their impact on surrounding agave populations could be more easily quantified.

Worse yet, corporate-backed distillers simply hired otherwise unemployed workers or released convicts to go into the *monte* wildlands to destructively harvest every near-mature plant they could find. While most well-established *vinatas* of traditional Sonoran families would not allow this kind of *saqueo* or pillage to occur on their lands, there were many disheartening reports of illegal, clandestine harvests occurring on *ejido* or collective rangelands, and in *reservas* or protected areas.

Curiously, the clarion call about overharvesting that reached politician's ears did not come directly from the *mezcaleros* themselves—although many grumbled to their family members about the rates of loss—but from bat biologists and pollinator conservationists. Around 1980, bat biologist Donna Howell contacted Gary and asked to be introduced to any of his friends known to be "moonshining" agave liquor just south of the border from where she worked with bat pollination of agaves in the Chiricahua Mountains of far southeastern Arizona.

Dr. Howell hypothesized that without bats, the pollination success of *Agave palmeri* had dropped precipitously in some places, suggesting simultaneous declines in bat populations and plant reproduction. That is why she wanted her documentary set in northern Sonora where "moonshining" of agave liquor was still a popular industry.

To prove her point, Donna showed us how she had been opening the seedpods of *Agave palmeri*—or Sonoran *lechuguilla del norte*—in the arid mountain ranges where she had been tracking bat populations for several years. When she counted the black markings on the inside of the pods, which indicated how many fertile seeds had been produced, the seed count was astonishingly low, indicating that bats may not have visited the flowers, only honeybees.

But when Donna took us into an herbarium of dried plant specimens and counted the black striations in seed pods taken from the same mountain range from early decades, the seed set was very high. She estimated that the current populations of this agave were realizing only 1/3000 of the sexual reproductive potential they had demonstrated in previous decades, despite some pollinations by bees and hummingbirds!

Donna became alarmed by how dramatically the bat-agave symbiosis was being disrupted, and alerted Gary and many others to the risks they faced.

Gary guided Donna and her film crew to meet a *mochomo* in the Río Moctezuma, not far from where Donna had accomplished her field studies in the Chiricahuas. The *maestro* who agreed to be filmed, Joaquín Martínez, nonchalantly estimated that he cut four to five hundred plants annually to make *mezcal lechuguilla* and *mezcal bacanora*. He was unabashed in his pronouncement that others were cutting far more agaves, often without letting them "pup" or vegetatively propagate to sustain the population.

Joaquín's willingness to talk openly about this issue was not only honorable but signaled the seriousness of the current dilemma. In the old days of Prohibition, if he had publicly acknowledged that he ran his own still, he might have been put in jail for three to four days without food. He would have been required to pay a fine and to give up all his bottles of mezcal to the sheriff.

But then, regulations had been locally relaxed by sheriffs more favorable to the tradition, and Joaquín and fifteen other *mezcaleros* were making all they could. Because of overharvesting by some among them, he was then forced to go as far as 45 miles (70 kilometers) to get his plants.

How far did the bats in that area need to go to get enough pollen and nectar to survive? We don't know exactly, but when agaves and columnar cacti are scarce, we estimated that they may travel as much as 120 miles (200 kilometers) in a single night to seek out abundant nectar sources with safe caves for roosting nearby them. Traveling distances may increase in landscapes where Sonoran harvesters have cut at least three species of paniculate agaves immediately before they bloomed. In the one area of Sonora we investigated, at least 7,500 agaves were cut and kept from producing the flowers required for the bat's nutrition.

When her controversial film *The Bat, the Blossom, and the Biologist* was released for international distribution in 1983, Dr. Donna Howell offered this comment: "This swath of foodlessness is responsible for the well-documented decimation of populations of nectar bats, especially over the last thirty-five years." Declining seed set, as demonstrated by our field work and Donna's meticulous review of historic herbarium specimens of agaves, clearly demonstrated a decline parallel to the decline in bat populations.

※

Among bat biologists and pollinator conservationists, Howell became a Paul Revere on a midnight ride to alert others of impending danger. There would be consequences to disrupting the mutualisms or symbioses between bats and agaves—or agaves and mezcal drinkers. If we want to drink without guilt, we cannot continue to overharvest agaves without consequences for their attendant wildlife.

Her seminal work inspired others, and collectively, their results slowly began to change everything in the world of agave conservation and mezcal production. It led to the Forgotten Pollinators campaign that Gary co-founded with ecologist Stephen Buchmann, and then to the multi-million-dollar Migratory Pollinators initiative that allowed him to pass much-needed funding support to the likes of stellar bat conservationists Rodrigo Medellín, Ted Fleming, Merlin Tuttle, and Karen Krebbs. In addition to a surge of funds to help with monitoring the status on bats all up and down their migratory corridors, Gary and his allies suddenly had face time with the highest-ranking environment policymakers in both the US and Mexico. They mobilized far more environmental professionals to protect bats and agaves than he could have ever imagined.

※

The remarkable work of the Tequila Interchange Project soon followed, demonstrating that some mezcal and tequila producers were fully willing and able to work collaboratively to restore bat-agave mutualisms in their local landscapes for the benefit of all. The migration of lesser long-nosed bats may cover 750 to 930 miles (1,200 to 1,500 kilometers) or more through a variety of private and public lands that harbor many kinds of habitats and have been placed under many different kinds of vegetation management. That meant that even the good work of individual diligent *mezcaleros* like Emilio Vieyra Rangel might not be sufficient to move the needle unless other land managers were recruited to reinforce his efforts. It would take entire villages, strung like beads along the same "nectar corridor" traveled by migrating bats.

Fortunately, many additional private and public initiatives began percolating up to plant agaves and protect bat roosts from Michoacán northward to Arizona and Sonora. The discovery of formerly unknown roosts for lesser long-nosed bats, combined with the onset of sizable reforestation efforts with agaves, was sufficient for the international community to make this agave pollinator the first bat species ever taken off the list of critically endangered wildlife in 2018.

As Rodrigo the Bat Man himself prophesized nearly two decades ago, "There is more than enough work to be carried out in the foreseeable future in the bat and agave conservation arena in Mexico. We have taken the first steps, and now the program requires solid strategic planning, numerous participants, and a dependable collaborative scheme. Never have the circumstances and situations been more promising to achieve bat and agave conservation; never will it be more feasible and timelier than now."

⁂

Chapter Thirteen

AN AGAVE AGRICULTURE

As we travel the Americas and to arid subtropical landscapes on other continents, we are always amazed to see agaves growing in some of the most elegantly austere places on the planet. While much of the scrappy shrubs and wildflowers in these landscapes hunker down against the ground to spread their roots into every nook and cranny of the dry earth, century plants seem to rise above them. With their winglike leaves spiraling around their hearts, it sometimes seems that they are ready to soar into higher places, as dragonflies and helicopters do whenever they catch a dry air thermal and lift off the ground with the greatest of ease.

In our minds, agaves are ready to soar into new roles that could potentially be of great service in this time of rapid climate change. With the demand for mezcal and other agave distillates set to multiply three to seven times within the next decade, and many wild agave populations showing signs of decline, what kinds of agave agricultural production might help us rise to meet these challenges? Can we halt the losses?

You might be surprised to learn that the kinds of novel agricultural habitats we have in mind for Mezcalandia will not look much like Tequilandia, where row after row of sword-shaped, powder-blue leaves cover the ground in endlessly repeated patterns, with nothing between them and the equally

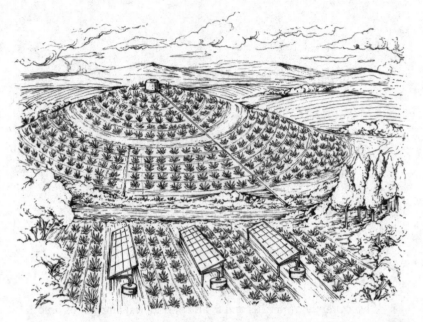

Farm of future with agaves alley-cropped with agrivoltaics

blue sky. And the many mezcals produced in the future will not taste or feel anything like the tequilas of the last half century of boom and bust. Let's try to envision just a few of the new roles agaves may take on as we move into a Braised New World.

Close your eyes and imagine another kind of future for our favorite beverage plants. Open them again. You may glimpse a scenario where succulent rosettes of gray, green, and blue agaves grow together, partially shaded by the tree canopies of *mezquite, huisache, bonete,* or *guamúchil,* alongside arborescent prickly pears, vining dragon fruit, and towering organ pipe cacti.

On the remote chance that abundant rains arrive as they used to, ground-hugging, nitrogen-fixing legume vines will spread across the earth below the up-curving leaves of agaves, and farmers will sow nutritious, heat-tolerant seed plants like grain amaranths, chia, chickpeas, salt-resistant *romerillo*, quinoa-like *huazontle*, red flax, sesame, tepary beans, and purslane or *verdolagas*.

A variety of tree crops in agroforestry plantations or in hedgerows on the field's edge will offer all the needed fuelwood for making mezcal from the rows of agaves that they flank. The trimmed leaves of agaves will be shredded, fermented, and mixed with *mezquite* pod flour to feed goats, sheep, quail, or chickens. The iconic *milpa* landscapes of Mesoamerica will be exemplified further and anew, with agaves interspersed among the many races of maize, beans, and squashes.

In such a future, agaves will be cultivated and allowed to mature for flowering in quantities large enough to take the heat off wild populations, but they will never be grown in disease-prone monoculture again. And this polyculture of agaves, cacti, and hardy tree crops will pull down far more carbon to slow climate change than can any kind of annual crop field in the world.

※

Sound too fanciful, too far out, and too far off? Viable models for desert agroforestry already exist in Arizona, Sonora, the Canaries, and the desert oases of North Africa. But nowhere else in the world do models of agaves nested in a slow-growing polyculture of perennial food and beverage crops already seem as promising—and as profitable—as they do in Techaluta de Montenegro, Jalisco, and on Rancho Vía Orgánica, outside San Miguel de Allende, Guanajuato.

One sunny January day, we drove through the grassy flats and salt-lined playas twenty minutes west of Lake Chapala, where thousands of migratory birds were congregating for the winter. But after we turned off the highway and started going up volcanic ridges toward the village of Techaluta de Montenegro, the landscape become more agrarian, and its habitats more heterogeneous.

When we decided to pull over and get out of the vehicle to look around, our companions were not exactly sure what they were seeing. This patchwork quilt of succulent plants and trees seemed so peculiar because the mix of plants we were witnessing featured many kinds of crops, not just the two kinds of cultivated magueys so often destined for mezcal distilleries.

These spatial patterns and forms of crop plants simply did not fit the ste-
reotypes of the "conventional" modern agriculture at all. There were curvi-
linear rows of agaves next to other rows of tall cacti. Native corn, beans, and
squash were "alley cropped" between these thorny succulents in 10- to 15-foot
(3- to 5-meter) swaths. There were also dragon fruit cacti perched or vining
up on the trunks of deep-rooted trees that provided partial shade from both
the edges of these fields and in the very heart of them. These woody perennials included *mezquite*, with their sugar pods; *guamúchil*, with their beautiful
hardwood trunks; and *bonete* trees, a papaya relative that offers large edible
fruit with delicious red and yellow flesh.

In the oldest trees, the vine-like stems of dragon fruit cactus sprouted
straight out of the elbows and crotches of these woody elders, even where
there seemed to be no wind-blown soil deposited to shelter their roots. And
yet, the very next summer when some of us visited Techaluta once more, we
witnessed farmers on ladders harvesting their high-priced fruit for the most
elite markets in the burgeoning metropolis of Guadalajara.

Yes, this "agroforestry," "intercrop," or "polyculture" of various lifeforms
might have bewildered any conventional farmer. It looks chaotic, or even
unproductive to any American or Canadian tourist who got lost on their way
back from Lake Chapala's high-end craft markets, or from those similar
attractions in the Tlaquepaque district of the capital city of Jalisco.

Unruly, maybe, but not unproductive. Techaluta's farmers are involved
in a supremely productive, resilient, and lucrative agricultural endeavor that
offers delicious food and beverage products to their own community, with
their surplus now sold in Guadalajara and Los Angeles.

❧

The four thousand members of farming families in Techaluta gain respect-
able incomes by selling thousands of fresh, high-priced fruits into specialty
markets and by offering their agaves to artisanal distillers of mezcal in their
area. Techaluta's most entrepreneurial women prepare value-added *ponches*,
jellies, and sauces from their family harvests to sell at roadside stands and
festivals in their area. And their large, healthy agaves demand prices from
distillers much higher than what the cloned *tequilana azul* plants can garner.

Eric Toensmeier, who works on agricultural solutions to climate change as senior fellow with Project Drawdown, confirms that perennial polycultures like those in Techaluta typically yield far more food and drink and pull down more carbon per planted area than monocultures of agave or other crops grown in isolation.

Sensing that same possibility, Mexican ecologist Veronica Zamora-Gutiérréz has worked with the farmers in Techaluta, gaining their respect and affection. She is joined at times by a team of British and Australian zoologists, pollination ecologists, and economists. They come into the mixed *potreros* and wildlands of Techaluta to help her evaluate the role of nectar-feeding bats in these food-producing landscapes. What they have discovered through their collaborations with Techaluta farmers is that taking care of the bat pollinators of the region by providing them access to the blossoms of additional agaves and cacti in their fields considerably increased the value of their harvests.

How might that be? For starters, keep in mind that as farmers in Techaluta historically found that there was commercial demand for the agaves and cacti that grew on the volcanic slopes above their fields, they brought them into cultivation to increase the density, productivity, and care given to these resources. That no doubt attracted and sustained a larger bat population for there was then more protein-rich pollen and energy-rich nectar to nourish them at this way station along their migration route.

But what escaped the eyes of agricultural scientists for many years was that the bats enhanced the value of the farmers' harvests through their effective pollination services. With the fruit-producing cacti found in Techaluta, greater frequency of pollination by bats (as opposed to bees or hummers) increased the fruit set, fruit size, and Brix content of soluble sugars in the delicious pulp and juices of pitaya and dragon fruit.

When the bats were excluded from the organ pipe cactus blossoms in a way that allowed other pollinators—bees and birds—to enter, the yield of their fruit dropped by 35 percent, their size and weights nearly halved, their shelf-life declined, and they were 13 percent less sweet than the bat-pollinated ones. The farmers not only harvested more fruit when bats were encouraged to work the blossoms, but the fruit were of such higher quality that they garnered higher prices in the marketplace.

We might argue that a similar but more subtle trend of adding value to agaves may also be occurring. Because the agaves grown in Techaluta often come from seeds of open-pollinated wild or cultivated plants, they can potentially express more hybrid vigor and variability. They garner prices in Tapatio nursery trade far higher per plant than those of cloned *tequilana azul* propagules. They likely embody more flavor and aroma in the agave spirits if the plants are allowed to mature longer than the precocious plants of *tequilana azul* clones. They are also likely to be less vulnerable to pestilence and plagues if grown in mixed-age, mixed-variety, mixed-crop polycultures.

In the rural communities of Jalisco, San Luis Potosi, Durango, Puebla, and Oaxaca, these contemporary designs or *trazos* of agroforestry plantations echo some of the most enduring forms of agriculture that once dominated both Mesoamerica and Arid America. Elements of those ancient food-producing systems are now being revived to generate a "climate-friendly" slow agriculture that may help future farmers in Mexico and the US Southwest endure the challenges of climate change.

A slow agriculture is one that provides yield stability of food and beverage crop from year to year without overtaxing its resources. Agaves being grown for mezcal in agroforestry intercrops are perhaps the finest example of such agricultural systems and their ultimate contribution to producing unique spirits. The many mezcal species, the firewood needed to roast them, and the herbs and fruits used to infuse them are all produced in the same polyculture.

Nowhere is this patient movement for climate-friendly agave farming more enlivening than at Rancho Via Orgánica, a regenerative agriculture and rangeland restoration training center less than a half hour from San Miguel de Allende in the state of Guanajuato. There, Ronnie Cummins, Juan Flores, and Gerardo Ruiz Smith have been using agaves and their companion plants for food, forage, fuel, fermented drinks, and carbon sequestration over the last few years.

When we visited Rancho Vía Orgánica in April 2021, we were amazed by how beautiful, healthful, and productive a well-managed polyculture can be. Agaves were everywhere, but so were *mezquite*, pomegranates, goats, and gorgeous vegetable gardens. The fields, orchards, and pastures on Vía Orgánica are so prodigious that seventy staff members, interns, and visitors can be nutritiously fed off their seasonal bounty year-around.

That is because every building has a cistern or water storage reservoir that holds all the rainwater capture off arable land surfaces, parking areas, roofs, and walkways. This is a farm in a semi-arid landscape at a 4,000-foot (1,300-meter) elevation that requires no groundwater pumping from wells, since the cisterns and soils rich in organic matter hold all that is needed. Vía Orgánica's cisterns remained full and its crops sufficiently watered even as 87.6 percent of Mexican croplands fell into the perils of lingering drought during that same year.

As if running a 200-acre (80-hectare) experimental ranch and farm as a regenerative agriculture training center is not enough, co-founder Ronnie Cummins has worked to build a larger coalition that has launched the Billion Agave Project across Mexico. The goal of this initiative by the non-profit Regeneration International is to plant 1 billion agaves globally to draw down and store 1 billion tons of climate-destabilizing CO_2. And as Ronnie reminded us several times during our stay with him, transplanting out agaves in moderately high densities on just 1.1 percent of Mexico's land area could completely offset all the carbon emissions in the entire republic.

We later walked out with Ronnie, agronomist Juan Flores, and permaculturist Gerardo Ruíz to a livestock pen where a herd of goats had just come in to rest and feed. Interns were mixing the *mezquite* meal in with the fermented agave leaves. As soon as the interns dumped the mixed silage into the feeding troughs, the guard dogs for the herd ran out ahead of the goats to get first dibs at the tasty silage. The goats soon followed, crowding around the trough like it was a magnet until all the silage was consumed.

As we walked several hundred more yards to where agave leaves were being fed into a customized chipper-shredder, Ronnie continued to explain the potential of the agave-based polyculture to lift low-income families out of grinding poverty. He bubbled over with excitement as he explained its multiple benefits to the poorest of families in the valley.

"Alfalfa and other purchased hays are extremely costly here, so the herders cannot afford them. . . . They try to make do with the little grass and herbs that come up in their run-down pulque plantations. But now, in a semi-arid landscape . . . that historically suffered from a lot of overgrazing by cattle, goats, and sheep, we now we have an agroforestry system with agaves and *mezquite* that reduces the pressure to overgraze brittle rangelands, while improving soil health and water retention. The livestock—particularly sheep and goats—gain more weight on our silage of agave and *mezquite*, which produces tastier meat. The higher market prices they garner then can give the herders a profit margin, rather than driving them into further debt."

Best of all, the agaves and their *mezquite* nurse trees are drawing down and storing massive amounts of atmospheric CO_2 to slow or mitigate climate change.

But how do they pull that off (or down)? Between sips of fruit-flavored pulque, Ronnie scribbled numbers on a paper napkin, then walked us through his calculations. According to Ronnie's and Juan's calculations, once a system has two thousand agaves alley-cropped with five hundred trees like *mezquite*, mulberry, and pomegranate, the agaves themselves can sequester hundreds of tons of carbon, even as 20 percent of the agaves are harvested annually. When we consider the carbon pull-down by the tree species in the system, it achieves some of the best carbon sequestration values for any cropping system in the world—easily 500 metric tons of carbon per hectare, in a desert polyculture.

If the value of buffering farmers from climate change is not enough of a reason for them to intercrop agaves, Ronnie reminded us of all the revenue streams that come out of this multi-cropped system: aguamiel, pulque, mezcal, sweet sap syrup, honey, and edible *mezquite* flour, as well as prickly pear and other cactus fruits, mulberries, pomegranates, firewood, goat, mutton, and beef. Before his untimely death in 2023, Ronnie convinced many that these new revenue streams from holistic management of intercropped agaves are enough to eliminate debt and to bring prosperity to formerly poor farmers:

"Combining the market value of the *penca* [fermented, trimmed leaves] and the *piña* [hearts] of the three most productive varieties of agave we grow at Vía Orgánica, we arrive at a total gross market value of $152,500 per hectare—that's $61,538 per acre—over ten years. Every year—starting in year two—we harvest 20 percent of the *pencas* from each agave plant, which then grow back and replace the biomass harvested to make fermented silage for livestock. The agave silage (wet, not dry weight) capacity in this perennial polyculture averages 100 tons per year, when estimated over ten years. And that amounts to one of the cheapest sources of feed for goats, sheep, and cows you can get in desert areas."

Many of Ronnie's claims are independently supported by other agave researchers. As one of the premier plant ecologists working in Mexico deserts, Exequiel Ezcurra cannot understate the promise that agaves can offer Mexican farmers facing hotter, drier growing conditions: "As succulent plants with CAM [crassulacean acid metabolism], agaves open great possibilities as new crops with high water-use efficiency, as dealing with water scarcity becomes an increasingly important problem in Mexican agriculture."

There are many reasons that we should immediately begin to employ more CAM crops like agaves in mixed cropping systems. We need only look as far as the predominant temperate crops (of the more water-intensive C3 metabolism) to see that agriculture in two-thirds of Mexico and the arid US West is broken. Hundreds of thousands of acres are falling out of production—150,000 arable acres in Arizona alone from 2021 to 2023—due to multiple stresses on farmers' bottom lines. The heat waves and droughts that have recently hit Sunbelt agriculture again and again have dramatically diminished yields while raising production costs. That is because most vegetables, legumes, and fruits have hit their thresholds for optimal growth and profitability.

We hate to sound like doomsdayers, but the situation is steadily getting worse. Due to recent water rationing mandated by pervasive drought in the Colorado River basin, farmers in Arizona alone have lost well over a million acre-feet of irrigation capacity since September 2021, prompting immediate losses of thousands of rural jobs and $200 million a year in farmgate sales. Top that off with the fact that farmworkers in desert states are facing an unprecedented risk of lost work time from dehydration, heatstroke, and fatigue-related accidents and you have a perfect storm (without any water!).

Let's look more broadly at what has happened with farming in an arc within the US-Mexico borderlands centered on Tucson, and ranging west to Mexicali and the Imperial Valley, or east to Las Cruces, El Paso, and Juarez. Many farming valleys in the Mojave, Sonoran, and Chihuahuan deserts have hit their lowest level of annual precipitation in the history of recorded weather data, receiving less than a fourth of the average rainfall these farmlands averaged over the last thirty years. Since 2020, the drying up of irrigation reservoirs along desert rivers has exacerbated the over-pumping of fossil groundwater with fossil fuel. This has not only raised farmers' energy costs of irrigation delivery to crops, but contributed to some of the highest greenhouse gas emissions of any farms in North America.

More than three-quarters of all farmlands and rangelands in the arid US West and Mexico have suffered from what was once considered to be "exceptional" heat and drought. These conditions are now the norm not the exception, forcing stockmen to seek out alfalfa for their cattle, sheep, and goats as supplemental feed. But growing an acre of alfalfa in desert regions takes as much as 9 acre-feet of water, roughly six times the volume that an acre of agaves requires. With once-perennial streams and springs remaining dry for 100 or more days straight for several years running, ranchers also need to haul water in trucks to provide drink for their livestock when they are most in need.

And yet, agaves seem to be surviving and even thriving in the Mexican and US states most affected by the perilous drought and heat loads attributed to global climate change. Nonetheless in the US, agaves have been used more for "urban beautification"—water-wise ornamental xeriscaping in parks and yards—than as food and beverage crops for climate-smart agriculture. That is unfortunate, for agave crops could have helped farmers survive recent climate calamities with far more resilience than the water-guzzling annual crops grown today. The agaves could have also drawn down enough carbon to offset the greenhouse gas emissions that have accelerated climate change.

Any one of these challenges—water rationing, rising energy costs, heat stress on crops and farmworkers—would be daunting enough in and of itself. Together, they suggest that the prevailing paradigm for conventional agriculture in desert regions, particularly in the arid US West and northern Mexico, is on the verge of collapse. But our federal agricultural agencies have been tardy in investing in long-term solutions to these challenges, and some states like Arizona have defaulted on developing drought contingency plans effective enough to stave off disasters. Most philanthropic foundations in the US are headquartered in more temperate climes within 20 miles (32 kilometers) of seashore or lakeshore, and seem barely aware of how much stress farmers are already suffering in the more arid hinterlands of the North American continent.

In Mexico, other than funding the Sembrando Vida reforestation workforce, the federal government has offered few incentives for the businesses involved in tequila and agave nectar production to return to growing multiple kinds of agaves in the same *potreros* with other crops. That lack of foresight might persist until the disincentives for maintaining blue tequila monoculture become so prevalent that its paradigm also fails. Then, Necessity will become the Angry Mother of Invention once again.

Fortunately, most of the traditional agave agriculture in Mexico—from pulque production in the semi-arid highlands south of Guadalajara and Monterrey to mezcal in the dry valleys from Mexico City to the Sierra Sur of Oaxaca—already favor multiple varieties and species that are intercropped with the traditional staples of Mesoamerica. Agaves are still grown within yards of maize, several species of beans, squashes, amaranths, tomatillos, chiles, and a dozen kinds of cultivated greens. These traditional polycultures have survived five centuries of suppression and oppression and are not likely to disappear soon. We await the renaissance of multi-cropped *metepantles* all across Mexico and the arid US West.

The looming dilemma is that most producers of agave spirits other than tequila are increasingly lured into adopting mechanized farming techniques that further increase the fuel and water use for more modernized agave production systems. Our hope is they may choose energy-efficient, affordable technologies that allow them to integrate best contemporary agroecological practices for harvesting multiple crops off the same parcel of land. In that manner, they can build upon—rather than reject—the *milpa* and *metepantle* traditions of agave terraces in Arid America and Mesoamerica. Small biofuel- or solar-charged tractors, scoop harvesters, and chipper-shredders can be skillfully employed to reduce tedious labor without damaging plants or soil.

Of course, we are not prescient enough to predict which direction most small-scale crop producers and distillers of artisanal and ancestral agave distillates will move: toward industrialization or toward agroecological innovation. Unfortunately, there are already some signs that the farming of *espadín* agaves in Oaxaca has gone too far in the direction of tequila's blue deserts.

In some stunning reporting by *El Universal*'s Juan Carlos Zavala, who is among the most astute journalists in Mexico, the scaled-up farming and distillation of *espadín* agaves has already been seriously depleting water resources, fuelwood crops, soils, and agaves within the larger valleys in his state. Zavala received state-level mezcal production data from Oaxacan official Elena Cuevas Hernández, a conscientious legislator from Morena, Oaxaca, capable of in-depth scientific analyses.

Cuevas's data suggests that to produce *every liter* of mezcal from farmed *espadín* agave in 2019, 20 liters of water and 6 kilos of firewood were required. That is eight times the water and nine times the amount of firewood used in the mezcal industry in 2011, when just a third of the volume of mezcal distilled in 2019 was being more traditionally produced on a smaller scale. Zavala's conclusion was that the cost of the mezcal industry getting bigger and more like tequila in its production costs is taking its toll on the Oaxacan landscapes and workforce.

While such trends remain worrisome to most of us, we are optimistic that the teetering microenterprises associated with most brands of mezcal and other agave distillates will right their ships and sail in a more sustainable and delicious direction.

Axel Visel, an expert on low-input crops for bioenergy production, has neatly summed up why these hardy succulent crops will be key elements of any climate-friendly polyculture in the future: "Agaves are capable of growing on marginal lands otherwise unavailable for food production. Agaves annually produce 1,034 metric tons of dry biomass per hectare. In some cases, their productivity can fall into a range comparable to biofuel feed stocks grown with higher water and nutrients. . . . The water use efficiency and thermo-tolerance of *Agave* species enable their cultivation on arid lands unsuitable for staple food production, allowing bioenergy production with reduced impacts on the environment and world food markets."

The water use efficiency that Visel alludes to—the amount of edible biomass for the volume of water used by the crop over a decade—can be four times greater than that of grains like corn or sorghum, and six times that of legume crops like beans, soy, or alfalfa. In short, the same amount of water will yield two to six times as much food or beverage if dedicated to agave than if dedicated to most legumes, grains, and forages.

The growing number of farmers and ranchers involved in Regeneration International's Billion Agave Project are not typically interested in planting their investing in thousands of magueys in monocropped rows that stretch unimpeded to the horizon, as is done for corn or soy. Like their prophet, Ronnie Cummins, they want diversity, nutritional density, and reduced greenhouse gas emissions that can mute the rate of climate change down to a dull roar. And the three "sassy sisters" of agave—prickly pear, cactus, and chiltepín—can offer more flavor, nutrients, and antioxidants with less water than the more-famous three sisters of corn, beans, and squash.

Other innovators are interested in growing plants for small-batch spirits suited to and celebrating the unique terroir of their own home place. They seek to bring their land to optimal conditions for the long-term cultivation of specific agaves, and its capacity to produce firewood and harvest water without overtaxing the landscape.

We are delighted that there are now hundreds of young *maestras* and *maestros* who are on fire about these possibilities, for they are cooking up a fresh vision for agave distillates. If they can join forces with the carbon farmers, they will have the means to turn away from potential danger into farming practices that are more regenerative and distillation practices that are more delightful to our eyes, lips, noses, and throats.

Chapter Fourteen

THE CREATIVITY OF
MAESTRAS MEZCALERAS

On our way toward the Sola de Vega valley in southwestern Oaxaca, we passed under a sign that declared "Mezcal is not some fashionable trend here, it's our tradition!"

Less than an hour later, when we met up with Maestra Sósima Olivera Aguilar, it seemed that she was the living embodiment of that *dicho*, that folk saying. And yet we hardly had time for small talk when we arrived at the *palenque* where she was working that day, for she immediately hustled us up from the Tres Colibrí Cooperative distillery into the fields on the ridges above it. That is where she wanted to show us both the agaves and firewood trees being grown for use at the *palenque*.

"Let's head uphill this way," she said, as she pointed to our destination with a machete of a length nearly half her own height. As she beelined it up the trail, she whacked at weeds and cleared away any obstacles that might have otherwise slowed our pace.

Somewhere ahead us, Sósima shouted out to the workers in the fields who remained beyond the reach of our eyesight. As her voice rose up from beneath her huge straw sombrero, they heard their names being called, as she cracked jokes and sang short stanzas of their favorite songs. They called back to her. While she sought out the men to introduce us, she whisked by a dozen kinds of maguey plants that were all taller than she is. Sósima makes up for her short stature with a far-reaching vision and passion for their collaborative efforts to change the ways mezcal is made and marketed.

"You know," Sósima said as we caught up to her, while she pointed to where the elderly farmer Juan de Jesus Díaz and his young helper stood amidst the agaves, "these skilled workers do not wish to bend to the market's demands. Instead, they feel that it is the market itself that needs to readapt to the needs of the agave, and to those of us who care for it. And if the US and Europe do not fully understand or accept what we are trying to do, we are more than happy to keep the best agave distillates here in our community, so that the people who understand how much sweat goes into every bottle can be the ones to savor and celebrate it!"

Sósima then walked over Señor Diaz and gave him a hug. She asked him to identify and describe the various agaves within reach of us, which he did with an expert's lexicon that captured all the heterogeneity we were witnessing. He took us through the details of leaf shape, color, and dentition; the number of years to maturity; and the capacity to produce seeds or vegetative offshoots for at least a dozen varieties of agave that grew in his field amidst maize and pumpkin vines: *arroqueño, barrileto, coyote, espadín, mexicanito chiquito, mexicano grande, mexicano penca larga, sierra negra,* and *tobalá.*

"You can plainly see," Sósima whispered to us as Diaz walked ahead to locate another subtly different variety of maguey, "this is more than just a business for him. For us, traditional mezcal is not just merchandise to sell; it is an essential element in our *cosmovision* . . . how do you say it, *worldview?* He is elaborating the kind of spirit that unites all of us here locally during times of trouble, times when we come to celebrate the successes of our collective effort, or times of sadder or more tragic moments of our workaday lives."

Sósima does not use the term *collective* in a loosely casual or superficial manner. She co-founded the Tres Colíbri ("Three Hummingbirds") Cooperative several years ago in her birthplace, the Chontl village of San Miguel Suchíltepec Yautepec. That fair trade collective now involves more than seventy mezcal producers, after it expanded in 2010 to the Zapotec village of Villa Sola de Vega, which is three and a half hours away from her home.

Sósima began her first fair trade co-op when she was barely fifteen to sell herb-infused mezcals and other artisanal products made by her grandmother and other Chontl women from San Miguel. Sósima was both inspired and encouraged to consider managing a fair-trade store by a pioneer in the Movimiento del Comercio Justo, former Bishop Arturo Lona Reyes, who coincidentally died the very day we first met Sósima outside of Oaxaca City. Known as the Bishop to the Poor, he was regularly seen at public events with simple huaraches on his feet, clad in blue jeans and a white T-shirt, with a big wooden cross on his chest. He was key to jump-starting many of the fair-trade coffee enterprises on the Isthmus of Tehuantepec near where Sósima spent her childhood.

Bishop Lona and his devotees also supported Indigenous youth like Sósima in efforts to start their own microenterprises based on the principles of fair trade. Because of his radical politics and fervent support for the Coalition of Workers, Peasants, and Students of the Isthmus (COCEI), Lona Reyes once suffered from an assassination attempt by right-wing vigilantes, and was forced to resign as bishop in 1998, just as Sósima's co-op was struggling to maintain its foothold in her community's economy.

Although she went off to college to receive formal training to hone her entrepreneurial leadership skills, Sósima decided to return to her family's deep agrarian roots, which were soaked in mezcal: "When I was a student in college and was deciding where to practice my newly learned skills and values, it suddenly dawned on me that I had always wanted to run a *palenque*, but not just because I desired to distill my own mezcal. I had grown up romping and roaming all around my grandparent's *palenque*. I thought it was paradise, true paradise! It was just above a spring that supplied water to the distillery, so I could run down into the water and splash around whenever it was hot. I could climb up into the branches of the fruit trees that shaded the fermentation tanks and pick guavas to my heart's content. Then I would swing down from the trees and chase the hens away from the roasted *piñas* of agaves. I would crawl around looking for where they had hidden their eggs."

Sósima continues, "When I was very little and my parents were helping my grandparents distill mezcal well into the evening, they would put me in a dried fermentation vat to nap where no mosquitoes would reach me. I fell asleep to the smells of the ferment in the wood. And whenever I caught a cold or suffered a sore throat, my mother—who was also a midwife—would give me a tablespoon full of herb-infused mezcal to take care of my cough."

Sósima giggles through her sentences, "How could I *not* return to the making of mezcal, given that my childhood playground was the distillery itself? That is where I cut my teeth. By that I mean that I first became tipsy after drinking a small *jicarita* cup of mezcal when I was only eight years old. It cured my cold, but it did not extinguish my fascination with fermentation and distillation. It is not at all surprising that after being educated, I craved to come back to the *palenque*, because it was such a formative, creative space for me."

While we thought that we would be immediately returning to the *palenque* with Sósima, she first invited us to go farther away, out past the edge of the cultivated agave fields. She had her own reason for doing so, and it echoed what we had learned from *mezcalero* Lalo Ángeles in Oaxaca many days before: "You can't have a good *palenque*—a well-run distillery that makes good mezcal—unless its products are rooted in well-managed healthy agriculture. And you can't have good agave agriculture if your *milpa* is isolated from wildlands. If there are no bats to pollinate your agave blossoms to produce open-pollinate seed for the next planting because all the natural vegetation near you has been bulldozed, your *palenque* will suffer. If the *desmonte* caused by the bulldozers eliminates that wild agave populations whose pollen could add variety to your agaves, you lose out. If you have to buy firewood from a half hour away because there is none left on your own lands to use in the distillery, your costs go up. That is because you must pay woodcutters for their labor and for the gasoline to drive the wood to you."

We said little as she led us over a ridge beyond the field's edge into subtropical thornscrub vegetation, for the beautiful setting stunned us. There she had planted fuelwood trees over an acre of wild vegetation, transplanting them in between existing trees. Farther on, she had planted *Agave potatorum* seedlings from the Tres Colibrí nursery stock taken from six different nearby locations.

We soon came upon a variant of *Agave potatorum*—the species locally known as *tobalá*—that was among the first to flower since outplanting five or six years ago. It was diminutive, but healthy and beautiful. Sósima grinned with pride that her efforts from several years before were already coming to fruition. This mature *tobalá* was already sending up its first flower stalk, amidst dozens of other species of wild plants that together formed a habitat healthy and diverse enough to attract bats, bees, and hummingbirds.

"I think it is fair to say that women are especially adept at thinking about the entire process from pollination to seed to outplanting to harvest to fermentation, distillation, and bottling," she told us. "Women see how we must aspire to gain quality at each of these steps, for that's what determines the quality of our mezcals and the livelihoods we generate."

As we walked back to the *palenque* to join in the crush of roasted mezcal on a new *tahona*, Sósima expressed a bit of irritation with the regulations enforced by the Consejo Regulador de Mezcal in Oaxaca: "There's nothing wrong with crushing the *piñas* of mezcal with the grindstone of the *tahona*, but we are now restricted to this method should we wish to call any of our products *mezcal ancestral*."

She frowned.

"*Fíjate* . . . Historically, most of the small family *palenques* around here didn't have a *tahona*, they used wooden mallets to crush the roasted agaves. We never used fancy words like *ancestral* or *artisanal* when I was growing up in my family's *palenque*. It was either traditional mezcal or it wasn't—like the stuff from Santiago de Matatlán made industrial style, with a worm thrown in for show."

"But now the Consejo Regulador is trying to force us into processes we never did traditionally, saying that we can't call it mezcal if we don't follow their recommendations. Outsiders are determining what we call our products that we have made just as our grandparents made it? We must disrupt all this nonsense . . . it's not getting us anywhere!"

Sósima Olivera is not the only innovative *maestra mezcalera* in Oaxaca. In fact, there have always been women involved in managing the mezcal supply chain. Ten to twenty elderly Oaxacan women remain famous for brokering their families' distilled spirits during periods when their sons, husbands, and brothers were not allowed to legally market distilled alcohol in their villages. These women were often *curanderas* (herbalists) or *parteras* (midwives) who offered herb-infused mezcal products in local markets, just as Sósima, her mother, and grandmother did in their Chontl community near the Isthmus region. Today, a number of *maestras mezcaleras* in Oaxaca have deservedly received national fame or international renown for their spirits, skills, and talents.

Full recognition of the women who anchor their families' efforts to produce agave spirits may be belated, but it is changing the public face of the industry in other states as well. Veteran *bacanora* producer Laura Espinosa came to our 2022 Agave Heritage Festival in Tucson to tell her personal story, along with three other spirited women entrepreneurs from other states. On a theater stage with large images of her family operation projected on a screen behind her, Laura recounted her journey toward engaging with the thirty Sonoran women already involved with various aspects of *bacanora* microenterprises. Some are distillers, others are marketers, while a few do every single task that their father or male collaborators do to move agave plants from nursery to field to still to consumer. Laura noted that while Sonoran women have handled critical roles in the agave spirit industry for some time, their roles remained hidden from those outside their pueblos until recently.

But Espinosa has been gratified to finally see a good number of professional women fully accepted in key roles in the *bacanora* industry, given that after forty-five years under the first Denomination of Origin for agave spirits, only 5 percent of tequila microdistilleries are in women's hands. The mezcal, *bacanora,* and *raicilla* industries are similarly skewed and may still have some lingering male biases, but the proportion of women engaged in agave spirits other than tequila is already much higher.

⚜

Among the most renowned woman innovators is Graciela Ángeles Carreño, a true force in protecting the integrity of agave spirits, for she sees a woman's patience and ancestral mezcal production beautifully aligned. Graciela reminds us—as her friend Sósima did as well—that mezcal alludes to culture, *cosmovision*, and religion, not just calories and commodities: "Mezcal is an element that allows us to communicate with higher beings, pivotal to the preservation and healing of the pueblo's collective soul."

When we arrived at the Mezcal Real Minero *palenque* one late October afternoon, just before All Soul's Day, we hardly expected to see Graciela herself, given the intensity with which Oaxacans prepare for Día de los Muertos. Her understudy, Catalina, greeted us at the gate to Real Minero's warehouse and walked us through their wondrously diverse botanical garden of agaves before we realized that Graciela was on site and would be joining us. We meandered around the warehouse along elegantly designed garden trails that took us through the various major groups of mezcal agaves that Graciela's LAM Project is bringing into cultivation and evaluation. But when we arrived in the nursery where thousands of agave seedlings and offshoots were being grown out under shade cloth in intensively planted beds of each variety, Graciela appeared out of nowhere to accompany us for the rest of the afternoon.

Speaking quietly and thoughtfully while moving gracefully from plant to plant, Graciela explained that Proyecto LAM is her philanthropic effort to move agave production into greater diversity and sustainability to honor the vision of her deceased father (LAM stands for Lorenzo Ángeles Mendoza). Its conservation and restoration efforts extend beyond caring for the agaves in great commercial demand to rare and poorly known species that have uses other than mezcal. It also embraces the propagation of four different tree species like *guamúchil* that have become locally depleted as the number of *palenques* in Santa Catarina Minas has grown exponentially over the last decade.

Graciela has a PhD in rural development from the Universidad Autónoma Metropolitana, but she also has developed a sophisticated knowledge of the taxonomy and horticulture related to agaves by bridging what she has learned from three generations of *mezcaleros* in her family before her with academic knowledge she gleaned from her time collaborating with university researchers. Slender, with long, black wavy hair that she pulls up when in the field, Graciela exudes a measured intelligence whenever she breaks her silence to say something. As we went through the nursery, species by species, she pointed out the agaves that few other *palenques* use for varietal mezcals.

One of them is the *tripón* cultivar of *Agave karwinskii*, the highly variable species that may have tall trunk-like stems, or elongated *piñas*. Another—seldom seen at all in Oaxaca—seemed to be a domesticated *Agave desmettiana* with long, beautiful leaves that lack sharp teeth along their edges. This prompted Graciela to note how stifling it is that regulatory commissions for different agave spirits mandate that only varieties on an approved list can be used. Trouble is, Graciela confided, the authorities don't quite get how scientific nomenclature serves a purpose separate from that of the names peasant farmers give to what they perceive as both useful and distinctive:

Linguists have recorded nearly six hundred names for different agaves in Spanish and the many Indigenous languages of Mexico. That is a lot for regulators sitting at their desks to wade through, let alone fully fathom. In addition, one common name can refer to distinctively different plants in the various regions of Oaxaca. For instance, a Nahuatl name like *teometl* might be used for four or five different species of agave. The same with *tobalá*, which *mezcaleros* might call *papalometl* or *papalome* elsewhere. The regulators often assume that these three names are used only for *Agave potatorum*, but one or more of them might be used for the rarer *Agave nussaviorum* in another part of Oaxaca.

Graciela asked us an insightful but impertinent question: "Why does a regulatory council assume it needs to be the one that solely determines what name is 'right,' when its members are neither producers nor botanists?"

From her warehouse and botanical garden, where the LAM Project is headquartered, Graciela drove us over to the Real Minero distillery, designed for her by one of her brothers, Edgar, an architect by trade. Edgar joined us in the tasting room, where we congratulated them both on the most elegant, clean, and efficiently laid-out *palenque* we had ever seen. The water and energy flows were especially well designed.

Graciela looked over at her brother and grinned. "This is how a *palenque* looks when there is both a woman in charge and an architect who have grown up in *palenques*. You know, we spend so much time here, we wanted our workplace to be both comfortable and beautiful so that everyone of our *palenquero* crew perceived what we do as dignified work."

While she distills in many more slow-cooking *ollas* (ten at a time) than we have seen in any other *palenque*, other parts of her process are streamlined for efficiency. Her fermentation tanks are relatively large—500 to 1,000 liters—and she combines the use of an automated chipper-shredder with the old-fashioned crush by wooden mallets. Again, Graciela expressed frustration with the regulations: "To call our spirits mezcal 'ancestral,' we would be required to crush the roasted agaves with the millstone of a *tahona*. But because of the modest size of the staff involved in our operation, we must start the crush of roasted agaves with a mechanical chipper-shredder, then use wooden mallets or *mazos* to complete the process before throwing it all into the fermentation tanks. For this sin alone, we fell outside the regulatory framework for marketing our mezcal as ancestral."

Despite such bumps along the way, the Real Minero distillery received certification by the Regulatory Council in 2004, just three years after her father put her in charge of the operation, while she was still in academic research part-time. In 2009, Real Minero submitted for and received a license to export its mezcals to Europe and the US. The first exports took place in 2010. On occasion, Graciela has had second thoughts about whether the investment to export was worth it.

As we sipped mezcal with Graciela and Edgar, she grew more reflective about her journey to change the face of the mezcal industry: "Our business model is deceptively simple. . . . We felt we deserved a place in the global market with other fine spirits but wanted to do so without damaging or endangering the quality of our products in any way. In my mind, tradition and modernization are not antithetical if we do not take modernization to mean industrialization. If we modernize certain aspects of our operations toward greater sustainability, we are ultimately adding quality to our mezcal and stability to our operations. If half of our mezcal is exported and the other half drunk with pleasure by the men and women of our own country, then we have made the value chain respond to our ideals rather than letting it undermine them. I have wanted to demonstrate that a *mezcalera* need not become a commodity herself but can change her position in the value chain to better fit with her own values and aspirations."

Needless to say, we are delighted to see so many innovative women taking up the charge to refresh the agave spirit world, and to bring their ethical sensibilities and creativity to bear on their industry. When David calls their vision one of the greatest paradigm shifts that has ever happened in the world of mezcals, he is not speaking in hyperbole. The old world of marketing tequilas simply looked like a hackneyed Tapatio scene from an old Charro cowboy film, where only men were at the table, demanding that pretty but silent señoritas bring them another tray of tequila shooters. The new vision is one where everyone has a place at the table, where women safeguard the integrity of the plantation—not just the final product—and where quality and ethics are part of the conversation as much as gross yield and price per bottle.

Chapter Fifteen

❧

AGAVES' ESSENTIAL
WORKERS

aps: Most of us love to gaze at them, or to run our index fingers down the routes that connect our most beloved places. Between the two of us, we have maps on our walls of mezcal production regions, the centers of origin for food and beverage crops, and the homelands of Mexico's many Indigenous cultures. Maps are not only good for documenting distributions and distance, they also trigger dreams of where we might go and what we might see.

But some maps also remind us of nightmares, of tyrannies and inequities. If any of us were to sit down at a table covered with maps that charted the primary areas of tequila and mezcal production in Mexico, to juxtapose them with others that show distributions of various levels of earned income, poverty, hunger, and levels of chronic disease, we might be horrified by what we see. In many (but not all) cases, the hot spots of rural poverty, high food insecurity, and pesticide toxicity are precisely where blue tequila and *espadín* agaves are agronomically produced and industrially distilled. Wealth has not trickled down to improve the lives of most workers in the industry.

For mezcal, this pattern includes farmworkers in eight states where more than 815,000 acres (330,000 hectares) of agave fields are located. Although just nine thousand of those rural residents are counted as mezcal "producers," the entire mezcal agave spirit supply chain in Mexico involves more than three hundred thousand farmworkers, pesticide applicators, woodcutters, distillers, and bottlers.

In Oaxaca—the third poorest state in Mexico—nine-tenths of all of Mexico's mezcal is produced. Despite the rise in income that mezcal is now generating in Oaxaca, the percentage of the rural population living in extreme poverty is nearly triple the national average. In other mezcal-producing states such as Guerrero and Puebla, the poverty levels are tragically higher, as they are in the *comiteco*-producing state of Chiapas and the henequen fiber–producing state of Yucatán. The income gained in poor Indigenous villages in Oaxaca and Puebla for cutting agaves and wood may be in the range of $65 to $450 per year, less than the official minimum wage in Mexico.

Two tequila-producing states, Jalisco and Guanajuato, are among the most violent in Mexico, as are the sotol-producing state of Chihuahua and the *bacanora*-producing state of Sonora. Michoacán, Guerrero, Querétaro, and Jalisco also have high levels of violence in their mezcal-producing municipalities. Street violence usually gets airplay, while slow death due to the hopelessness that poverty and chronic illness carry often remains out of sight and out of mind.

For far too long, the stench of poverty and pesticides have lingered in the air around *palenques*, *vinatas*, and *tabernas* even when you do not visually observe them. There, any benefits of agricultural modernization may have already been outweighed by the double burden of malnutrition and exposure to chemical contamination in an impoverished environment.

Increases in wages over time have simply not kept pace with rising food, energy, and health care costs among most men and women at work in commodifying agave spirits for foreign markets. They are in an altogether different dance than their ancestors were when they took care of the *milpas* and *metepantles* that were once dedicated to the Goddess of Maguey, Mayahuel. Yet many campesinos today continue with the agave traditions of their forefathers for cultural reasons, since their meager wages could fully never compensate for the amount of work they dedicate to all of the tasks required to make good mezcals.

The disparities and dysfunctions are even more evident in and around the UNESCO World Heritage Site called the Agave Landscape and Ancient Industrial Facilities of Tequila. Cruise down the main drag in any dusty rural town in Jalisco where tequila is being produced, and you will find much the same pattern of stores lining the thoroughfare. You will see cantinas and *depositos* for beer; mini marts for snacks, bottled water, and carbonated beverages; stores hawking cheap clothes; hardware and auto parts stores; and dozens upon dozens of outlets hawking insecticides, fungicides, or herbicides.

In full view of any passerby, it is obvious that the Dows and Monsantos have had a ubiquitous presence in the Mexican countryside over the last five decades, perhaps even more so than in the Land of Corn and Soy of the US Midwest.

While Mexico ranks fourth in the use of all manner of pesticides among Latin American nations, Jalisco is first among Mexican states in its intensity of use of such agrichemicals. Over the last quarter century, many of the billboards and storefront posters in the production zones for tequila and other crops tout the miraculous powers of toxins with names like Hercules, Marvel, Balazo (Bullet), Paraquat, Nuvacron, Furadan, Gramoxone, and Tordon.

However effective these agrichemicals may be in temporarily knocking back weeds, snout weevils, moth larvae, and root-knot nematodes, such toxins leave a hidden but indelible mark on the health of the harvesters and *jornaleros*. Many of these workers move between one agave plantation to another, but some of them also seasonally labor in avocado, maize, and bean fields as well. That makes it extremely hard to attribute their pesticide toxicity or cancer to their work with any single crop, or for any particular company.

According to a health survey of Jalisco farmworkers undertaken by the University of Guadalajara in 2005, 61 percent of those interviewed claimed that they had suffered pesticide poisoning at least once while on the job. Moreover, 39 percent of the farmworkers evaluated had high exposure to these agrochemicals, with repeated bouts of contamination and illness.

How can these farmworkers presume, let alone definitely confirm, that it was the handling of pesticides in tequila or maize fields that made them ill? Well-trained doctors appear in their rural health clinics only one in eight workdays, and the Red Cross urgent care clinics are typically staffed by paramedics, not physicians who know how and where to immediately order lab tests to verify which agrichemicals may be running rampant in a patient's bloodstream. For any of those who get sick while tending or harvesting agaves in industrialized fields, only a small proportion get seen by a physician with enough resources at hand to identify the specific toxin and dosage level to which they were exposed.

Correlations may or may not indicate causes. In most cases, the rural health care providers can only match the symptoms—dizziness, vomiting, headache, stomach pain, profuse sweating—with the known consequences of overexposure to certain highly toxic chemicals. Pinpointing whether the culprit was a carbofuran, chlorpyrifos, or glyphosate may be more difficult, since the worker may have exposure to multiple chemicals in the same work, and these may interact once absorbed into the body. Nevertheless, Mexico has recently mandated a total ban on glyphosates (such as Roundup) by 2024.

Recently, University of Guadalajara toxicologist Joel Salazar-Flores confirmed beyond a reasonable doubt that there are dangerously high levels of toxic carbofurans in the bloodstream of over one hundred campesinos living in Jalisco. To be clear, these affected workers labor not only on agave plantations but on farms with other field crops in the region as well.

And agrichemical exposure is not limited to those who toil in the fields. Significant levels of this toxin were also found in family members who did not work in the tequila fields, but did live adjacent to them. Several deaths in Salazar's study were tentatively linked to high exposure, although the level of precision in diagnosis was low.

Summarizing the results from blood samples from 117 Jalisciense campesinos who had worked in the tequila- and maize-producing Los Altos region of La Cienega, researchers Carlos Melgoza, Frida Valdivia, and Rodrigo Cervantes had this to report: "Substances such as Furadan, Paraquat, Gramoxone, Balazo T, and Faena run through the veins of a hundred farmers in the Cienega subregion of the Jalisco highlands; all these chemical compounds are recognized for their toxicity and high efficacy against pests and weeds in crops. According to data from the General Office of Epidemiology of the

Ministry of Health in the state, the chemical compounds of these products are responsible for at least one person being contaminated or poisoned every five days within one of the nine municipalities of the Cienega subregion. These substances may be contributing factors or causative agents to diseases such as cancer, diabetes, and hypertension."

The real tragedy of these illnesses and deaths is that both the University of Guadalajara and CIATEJ agronomists have determined effective means of reducing agave and corn crop losses from these pests and plagues without using such toxins at all. Instead, they recommend beneficial insect and microbes, bio-controls, organic pest repellents, and altered soil management practices.

Without a doubt, the emergency room visits by harvesters and *jornaleros* working in agave fields are not all linked to pesticide poisoning. Many of those seeking urgent care arrive suffering from deep wounds and excessive bleeding caused by their razor-sharp *coas* and machetes. Working under conditions that would fatigue most of us, the workers must chop away at a mature agave at least eighty times before the plant is trimmed and released from its roots. Most of us would need two hundred hits with the blade of a *coa*—not eighty—to do what a harvester does with the usually precise placement of his tools. But one bad hit that might glance off the plant is all that is needed to badly injure a worker's thigh, calf, foot, or hand and send him rushing to a clinic with blood spewing out of his flesh.

As climate change worsens, agave workers also suffer with increasing frequency from dehydration, heatstroke, muscle pain, backache, and exhaustion. And emotional health among these workers is nearly as serious as physical health. Many workers complain of anxiety, depression, and sleeplessness. A good number express worry about simply getting their families enough food to eat.

Over half of these Jaliscienses farmworkers tracked in health and economic surveys claim they are underpaid.

A farmworker named Vicente was asked if he could get by adequately on his current wages. He responded simply: "No, never, [because the pay] is not just. There have been people I have worked with who have been cut so badly on the job that their guts were dripping out of the wound in the abdomen [and they cannot afford to stop working to get to a clinic]. There are those that have suffered a scorpion bite and cannot get help. I myself have stumbled upon rattlesnakes in the field [that could have killed me]. We have been left to sleep in warehouses where rats run free all around us. It is not just."

Most harvesters are given daily wages or *jornadas* a bit above the levels offered to unskilled *jornaleros*, but unlike the *jornaleros*, they have protection from being laid off during periods when there is shortfall of agaves. The *jornaleros* hired for a day or week can go without any steady income for months at a time. The shortage of agaves available for planting and cutting in both tequila and mezcal fields has worsened the options accessible to harvesters and *jornaleros* over the last few years. Those who know only this profession often get the shaft whenever the industry faces wild fluctuations in demand.

Even when they can get regular work, harvesters and *jornaleros* are paid by weight for the already-trimmed agaves. A mature, intact agave weighs significantly more than a trimmed agave *piña*, so a worker must harvest more than double the field weight of an agave to make his wage, which is still remarkably low.

Most harvesters of tequila and mezcal agaves can cut only eighty to one hundred agaves a day with 6,500 to 8,000 hits to the plants with their *coas*—less, if they are in pursuit of wild agaves growing on steep terrain. For their precision and endurance, most *jornalero* day laborers in agave fields earn just one-twentieth of what they would make over the same number of hours as a farmworker in the US. As late as 2020, most highly skilled harvesters made just $20 to $30 for a full day involving seven hours of back-breaking work, and may need to cough up a tenth of that to a broker who books their labor with a plantation owner—or more, if contracted through a coyote.

You may hear that at the height of his prowess, a master harvester and crew boss gleans $120 per day. But such superachievers in the *jima* are few and far between, comprising less than 1 percent of the mezcal and tequila workforce. To gain such a rarified status, a single man must be able to cut about 140 mature agaves to render nearly 5 metric tons of *piñas* in a single seven-hour workday. Still, that may be as little as $17 per hour for most men and the few women in the workforce out in the fields, and few can sustain that level of effort day after day for more than a decade.

David has devised another way of thinking about how much a harvester's or *jornalero*'s work is valued. It assesses the harvester's work relative to others in the supply chain for tequila and mezcal. For their roles in cutting and trimming 1 ton of agave, a team or *cuadrilla* of six harvesters received $22.50 in 2022. Split six ways, that is just $3.75 for each harvester's portion of the work. The labor broker gets 10 percent of what the harvesters make, simply for gaining a labor contract with the plantation owner and providing local transport, water, and tools.

That ton of trimmed, mature agaves sent off to the distillery renders roughly 158 liters of distilled spirits. If that amount of agave spirits is parceled out into cocktails, like margaritas, that utilize about 2 ounces per cocktail, that amounts to 2,608 cocktails by David's reckoning—a low-ball estimate, by the way. So for about 90 cocktails per mature agave, that's roughly a $48 value at the bar or restaurant. But each harvester takes home only 0.16 of a penny of that value.

In a fancy bar in a US or European city, a cocktail with 100 percent agave spirits might sell for as much as $15 a glass, with the bartender often blessed with a $2.50 tip for his skill in fashioning the cocktail. But the agave harvester back at the Mexican plantation needs that bartender to sell 602 cocktails for him to get paid just $1.

This calculation has stopped us in our tracks, as it should stop you, too. It may make you suddenly toss onto the ground any tequila or mezcal remaining in your cups. But we would hope that rather than wasting the liquids of their labors, that you think (or pray) deeply before you choose to fill it up again.

If you are not enraged—or heartbroken, as we are—by how little is made by the harvesters who anchor the entire supply chain that brings us tequila, mezcal, or other agave spirits, perhaps you—like ourselves earlier in our careers—been naively drinking rather than thinking. But all of us can begin to support brands and collectives that are sincerely trying to change the dynamic.

We no longer need to be complicit in such human suffering. No interesting historic fact about mezcal, no cultural custom, no distinctive flavor or fragrance in its terroir justifies us to override the need to care for the health and well-being of the workers who bring us our daily drinks.

If it were only the plight of the harvesters that we have been ignoring, it would be one thing. But we know that far too many *cantineros* and bartenders are also underpaid. We know that far too many agaves are being extracted from the wild—far more than the numbers farmers, foresters, and restorationists are putting back into the ground. We also know that far too many drinkers are addicted to alcohol. They will go to bed with whatever spirit is beside them, no matter what its quality may be.

We have been humbly convinced that the entire way of producing and using the maguey must be transformed, not just one little link in the mezcal supply chain.

Yes, there are hundreds of thousands of other hands that go into the making of tequila and mezcal, and into pouring these spirits into *jicaritas* cups, shot glasses, or bulbous cocktail mugs. But those with their hands on the plants are often the ones in the entire chain who are likely to get the short end of the stick.

In fact, the Borgen Project reports that four-fifths of the Indigenous people in Mexico still survive on less than $2 a day. In Oaxaca, where tending agaves and making mezcal are the primary sources of cash income in nearly 60 percent of the rural areas, two-fifths of the Indigenous *mezcaleros* there still lacked potable water, half of their homes lacked electricity, nine-tenths lacked sewers, and more than half lacked reliable access to medical services.

All of that is deeply disturbing, but the promising news is that there are remarkable initiatives being taken to better benefit the health and well-being of harvesters, *jornaleros, maestros mezcaleros,* and their families. One such project is underway in the Oaxacan village of Santa Ana del Río, which even by Oaxacan standards is off the beaten track. Due in part to lack of access to markets and technical assistance, the average annual family income over the last decade has varied between $3,000 and $5,000. Although many of the farmers there produce mezcal on the side to try to supplement their income, a liter of their spirits locally sells for just 50 cents per liter.

In 2017, Heifer International offered to help the families in Santa Ana generate more sources of income and nutrition. The community chose four strategies that they hoped might raise them up from their nagging levels of poverty. The weather in Santa Ana has always been seasonally hot and dry, but now, with climate change, drought lingers longer and cuts into their yields of maize and beans. Total failures of annual crops have become more frequent in the arid landscapes of Oaxaca, prompting young men to seek wage work in the US, leaving home where *la tierra no da*—the land no longer gives. The loss of so many young men from Santa Ana del Río as climate refugees makes life doubly tough for their fathers and uncles, who must endure the burden of farm work on their own more and more. Heifer aspired to help Santa Ana break out of this tailspin.

First, the Heifer Mexico staff did what they are most widely known for doing: bringing in small livestock (laying hens) that could offer residents an additional source of income as well as more protein for family consumption at home. The women of Santa Ana began to tend flocks of laying hens that laid so many eggs that not all could be eaten locally. Heifer staff then helped arrange for a buyer who comes to collect the surplus eggs that the women offer.

Next, Heifer partnered with a UK nonprofit called Donkey Sanctuary to help maintain the health and comfort of the preferred beast of burden in Santa Ana. The donkeys in the village are now vaccinated, given medical care if wounded or injured, and fit with harnesses that ease their burdens when they are carrying the harvests or moving the *tahona* grindstones. As donkeys live longer, healthier lives, the cost of keeping them subsides and their value increases.

Heifer then went in an uncharted direction when it agreed to support the construction of greenhouses and shaded nurseries to increase the production of agave seedlings and offshoots. Bonifacio Cruz Ruiz was one of the first *mezcaleros* to construct his own backyard nursery, which produces thousands of new agave plants on trays for outplanting. These are now becoming available for reforestation and cultivation just as the mezcal boom had created scarcity.

Finally, Heifer helped the Santa Ana community of *mezcaleros* renegotiate to sell their mezcal for fifteen times more than what it was sold before—though the price is still a pittance. Regardless, within just three years of that increase, production more than doubled, which means that the nursery-grown agaves will be essential if they are to achieve sustainability over the long haul.

Wisely, the community chose not to put all its agaves into the mezcal basket. Santa Ana women have now developed value-added food items like agave "sweet sap" syrups and marmalades as by-products from their increased agave production. To their surprise, a container of syrup processed by the women returns more per liter to their families than do some varieties of mezcal made by their husbands or fathers.

While it is still too early to know whether eggs, healthy donkeys, agave nurseries, and value-added food products will be sufficient to pull Santa Ana out of its economic tailspin, there is hope that these collaborations will bear fruit.

※

Inulin-rich, slow-release sweeteners like agave sweet sap and agave-derived xylitol (a polyol sweetener or sugar alcohol) are becoming available exactly at the moment in history when Mexico desperately needs to get its epidemics of adult-onset diabetes and obesity under control. Diabetes is now the number one killer in Mexico, claiming eighty to one hundred thousand Mexican citizens' lives each year. And sadly, four out of every six Covid-related deaths in Mexico during the first months of the pandemic were of individuals who had long struggled with adult-onset diabetes.

About twenty million Mexican adults suffer from adult-onset diabetes, with a 48 percent increase occurring between 2006 and 2016. Over the last half century, the incidence of diabetes has doubled every decade. Most epidemiologists attribute this meteoric rise in diabetes to the fact that Mexico now leads Latin America in its consumption of highly processed fast foods abysmally high on the glycemic index. It has probably not helped that Walmart has become Mexico's largest food distributor and retailer since the beginning of this millennium.

And yet, the core solution to this problem is not far beyond the back door on most Mexican homes: the native, slow-release food plants that form the Mesoamerican Diet. According to a 2015 *Mezcal* book from Mexico's National Commission for the Development of Indigenous Pueblos, the many foods and fermented beverage products called mezcals can be part of the solution to human health and well-being throughout Mexico through their beneficial prebiotic and probiotic effects on a number of nutrition-driven diseases.

Roasted agave *mezontes* and *pencas* are high on the list of Mexico's most healthful foods, ranked high with prickly pear pads, *mezquite* flour, acorns, cactus fruit, chia, and a multitude of dry bean varieties. Xylitol sweeteners, incidentally, have 40 percent fewer calories than sugar, and are absorbed as complex carbohydrates (like agave inulins). As a slow-release food, they prevent blood sugar spikes in those who suffer from adult-onset diabetes so well that they ranked at just 7 on the glycemic index, compared to 65 for white cane sugar. Prebiotic maguey biomass from *Agave americana* has been processed into xylitol sweeteners with a little help from two "probiotic activists," strain CDS3 of *Pseudomonas* gram-negative bacteria, and strain 65S3 of the saprophytic genus *Bacillus*.

Imagine this: The annual cost to Mexican society of dealing with diabetes may be as high as $5.5 billion. Currently, that consumes approximately three-quarters of all Mexican government spending on health care, amounting to a $708 investment in each diagnosed diabetic person in Mexico per year.

While we have hopes that the many distilled mezcals never disappear from our tables and bars, we also wish to promote better health and well-being among farmworkers through a full return of the many foods and beverages historically produced from the same agave plants. Compared to the global acclaim now being given to agave distillates, the healthful fermented beverages and foods also called mezcal garner far less attention. As Mexican writer César Augusto Patrón Esquivel notes, "There exist very few texts that focus on the importance of agave products among the traditional fermented beverages and foods of Indigenous diets, on their cultural context, and on the way that they reinforce the ethnic identities of the original pueblos of Mexico."

We hope that this book will help correct this oversight, and assure that the many ways agaves have been eaten and imbibed once again contribute to the physical health, cultural richness, and economic well-being of those workers and communities that are most in need. It's not too late to keep Mexico's new motto from becoming "*Exceso en Calorias.*"

One way that is being done is by boarding distinctive agave products onto the international Slow Food Ark of Taste—a global compendium of rare and undervalued foods and beverages that can help rural communities restore and enhance their local economies.

One such project involves Mixtecan *mezcaleros* in Oaxaca with the recovery of the mezcal plants they call *Yavii ticunchi'i*. This Mixtec name refers to wild plant populations of *Agave nussaviorum,* a dwarf *tobalá* or *papalometl* agave with a range restricted to just four mountainous districts of Oaxaca: Coixtlahuaca, Huajuapan, Nochixtlán, and Teposcolula. This endemic agave grows in scattered patches on cliffs in Oaxacan sierras where pines, oaks, junipers, and spiny shrubs form open woodlands along cliffs in Oaxacan sierras. That's where the patchy ranges of its two forms reach from 5,000 feet (1,600 meters) all the way up to 7,800 feet (2,500 meters) in elevation.

As Slow Food promotes the conservation and traditional use of this rare agave, it is assuring that the nutritional and medicinal benefits of this mezcal tradition benefit its Miztec harvesters. In addition to occasionally being used in a fermented drink called *ticunche* or in agave distillates, the Mixtec use the innermost flesh of leaf bases and flower stalks to create a sour-sweet food called *yahui ndodo.*

To make *yahui ndodo*, the young stalks are baked in clay pots or earthen ovens, then mixed with a wild clover called *coyule* that imparts the *agridulce* flavor. Surprisingly, this sweet, candy-like food has anti-diabetic properties. In addition, the blossoms of this agave—called *cacayas* or *tibilos* in Oaxaca— are sautéed or stewed, then mixed with masa dough for making tortillas.

But it is the juices pressed from the leaves of this rare agave that have the most significant medicinal value in helping Mixtecans deal with circulatory problems, inflammation, and diabetes. Curiously, it is used as an herbal medicine for both humans and their livestock that have been injured in falls or accidents.

Esteemed for its healing powers, *Yavii ticunchi'i* is considered a sacred agave among the Mixtecs for ritual uses connected to the game *pelota*. There are also medicinal and ritual practices associated with the very special mezcal distilled from this agave, known variously as *el del patrón*, *el del cura*, or *mezcal ticushi*.

Over the last decade, it has become increasingly hard to get ahold of these rare plants, for they grow high up in the mountains. The traditional uses have not been major factors contributing to the rarity of *Agave nussaviorum*, as the Mixtec customs of preparing *Yavii ticunchi'i* for food, drink, ritual, and medicine are dying out. Instead, it has been the loss or degradation of its wild habitats affected by poorly controlled grazing that has triggered the decline of some populations of this rare agave. That is where the Ark of Taste process comes in: to promote not only the propagation but the renewal of traditional uses of this rare agave for food and medicine in its Mixtec communities of origin.

Other agave-related products contributed by Mexico to the Ark of Taste include a liquor made from the flower stalks of maguey, the edible white larvae found in maguey plants, and the red *chinicuil* agave larvae. By having an agave product boarded onto the Ark of Taste, producers aspire to create a distinctive niche for their brand, just as some livestock producers prefer to label their meats as grass-fed, free-range, hormone-free, organic, biodynamic, or certified naturally raised.

It remains to be seen whether having standing on the Ark of Taste in and of itself increases demand over the long haul in a manner that assures that the *jimador* or *jornalero*, woodcutter, roaster, distiller, bottler, labeler, brander, and bartender are all better compensated for their essential work.

That is why David has chosen another, out-of-the-box path for trying to improve the economic well-being and health of the field workers involved in Siembra Spirits. He simply asks a local community of workers what they need most. If it is an infirmary or urgent care clinic for "wounded veterans of the agave wars," he tries to find funds—personal and governmental—to make it a reality.

If it is a better teacher and more computers for the *jimador*'s school-age kids, David funds their schools. If it is acquiring some uncontaminated croplands near their homes where they can grow their own vegetables without the fear that they are laden with pesticides, he helps them build out that option. If it is a solar pump on a well that can provide them with fresh drinking water, he works with the community to build one.

Fortunately, David is not alone in his innovative efforts to relink the value chain so that it does not benefit just a few at the expense of all others. There is now a good number of brilliant and compassionate innovators—whom we call *whole drink system thinkers*—who want to redress the imbalances in the mezcal supply chain. And, of course, there are many fair-trade collectives and cooperatives that have emerged for agave spirits. There is even a women's bottling collective for small-scale mezcal distilleries in Santiago de Matatlán.

To be sure, with this infusion of new energy and ethics spilling in from many states and cultures, the mezcal supply chain will never again be as uniform as it has been for the last three decades.

And yet, as the writer and historian Domingo García has warned, mezcal remains caught between two competing ideologies of economic structure: the campesino concept of community economic well-being and the capitalistic concept of a commodified, extractive economy. Most collectives and cooperatives are trying to hybridize the two. But as evolutionary history teaches us, very few hybrids ever survive to regenerate forms of their own kind. The mule is loping along on an evolutionary dead-end street. Only time will tell which hybrid mezcal marketing models will survive, and which will go the way of Ligers, Tigons, Zonkeys, and Geep.

Chapter Sixteen

❧

EMPOWERING BARTENDERS

Somewhere between the wild of Mexico's best agave landscapes and the air-conditioned boardrooms of spirits conglomerates stands perhaps the only group with the success or failure of those models squarely in their hands: bartenders, known in Mexico as *cantineros*.

On a brisk January morning in Jalisco, before most restaurants in town had opened, sixty people arrived at La Tequila restaurant on Avenida México, in Guadalajara's Colonia Terranova. Many had probably worked until midnight the evening before, but there they were—bright-faced and neatly dressed, with carefully coifed hair and a warm air of camaraderie in their conversations. They had come together out of curiosity, and perhaps out of the growing commitment that they could collectively change the agave spirit world for the better.

David had known some of them for years but had also drawn upon the suggestions of the local restaurant owners and chefs that are among his old friends and neighbors. They pitched to David the names of prominent mixologists, innovative hipster-activists, and expressive raconteurs in the Metro area to invite.

Gary had been coming to La Tequila and several other acclaimed restaurants in Guadalajara for over forty years, but this was the first time he had met the best and brightest of their *cantineros*. After welcoming all to have some well-deserved coffee and to join in the morning's conversation, David explained why he'd asked them there: "You are the minds, voices, and hearts that guide your customers toward the best mezcals and tequilas that Mexico has to offer the world. You are the storytellers who guide drinkers into new adventures. But in a way, you are also the conscience of our industry—those who can warn away your friends and clients from more problematic spirits, or who can guide them to those made ethically and traditionally in ways we can all celebrate without any regrets. But to stimulate your thoughts on these themes, I would like to introduce you to some innovators who will speak to us in the initial roundtable of the morning. I'd like to invite four of our guests from the US—some with years of experience living and working in Mexico—to the front of the room."

David then called out a few names, and four of us meandered through the maze of chairs to reach the front of the barroom:

Monique Huston, the winningly articulate Vice President for Wholesale Spirits at Winebow in Chicago, who assembled one of the largest collections of scotches and whiskeys at Dundee Dell in Nebraska before branching out to embrace the terroir-rich agave spirits of Mexico. Her work has been featured in the *Wall Street Journal, Forbes*, and *Whisky Advocate*.

Francisco Terrazas—currently managing the Tucson Agave Heritage Festival—has hosted the Agave Sessions podcast. Francisco knows the back roads to the distilleries of Oaxaca as well as those in his Sonoran Desert homeland.

Michael Rubel, General Manager at Estereo in Chicago, a master craftsman and mixologist of Latin-infused cocktails, whose drinks have been featured at Lone Wolf and Violet Hour.

And Gary, who batted last.

Not surprisingly, all the roundtable participants deflected any attention directed toward them back to the bartenders themselves as mentors and field guides into the agave spirit world. They are on the front line night after night, not only in service to their clients but in service to the *mezcaleros* themselves. They are the translators, mediators, and key links in the spirit supply chain.

All the roundtable speakers reiterated the same message: We can have a healthy spirits industry only if those on the far sides of the supply chain—*jimador* and *cantinero*—are socially valued, empowered, paid adequately for their work, and kept healthy.

You could feel the appreciation that these messages generated rippling out like waves across the room—smiles, tears, and beaming faces.

"Sometimes we feel isolated, overworked, and undervalued," one young woman said. "Your recognition of our work—and your encouragement that we should speak up for what we believe in—is good news that we needed to hear."

Since David and his colleagues established Siembra Spirits in 2005 as part of Suro International Imports, its defining feature has been the collaborative relationship with bartenders in both the US and Mexico. Then, when he incorporated the Tequila Interchange Project into his portfolio of activities in 2015, he aimed to foster the power of bartenders as they came into face-to-face contact with distillers and agave researchers. The innovation that emerged from these dialogues at the onset of the project propelled the Tequila Interchange Project in new directions, as they continue to do today.

From the singular position they hold between the hidden treasures behind the bar and the patrons who come forward to enjoy them, bartenders are pivotal to the entire set of relationships surrounding all spirits. As New York bartending legend Phil Ward once acknowledged, "I have always told people that my backbar was my integrity. If I couldn't sell you what was on my backbar with a straight face, I wasn't doing my job right."

Of course, a skeptic might say cynically that Phil is just one of a half million bartenders in the US alone, and not all of them have yet risen to that level of integrity. And yet the best of them—if not the majority of them—are consummate storytellers. They are tasked with linking the willingness of their curious clients to consume novel drinks with the willingness of distillers to provide something distinctive to savor. To help those eager to experience some spirits that may be altogether unprecedented for them, bartenders must know how to distill the drink's backstory down to its essence.

This ethos of sharing memorable backstories to delight and inform is what drives both Siembra Spirits and the Tequila Interchange Project. They not only capture previously unforeseen insights about agave spirits, but they foster the willingness of each player in the supply chain to be a teacher. In essence, bartenders can be the gatekeepers of the many potential stories within a bottle of agave spirit. They can open up the possibility that spirits aficionados will seek out the stories of distillers, harvesters, and others in the supply chain as well.

Nevertheless, lumping the values of all bartenders into one sweeping catchall category makes no more sense than lumping all agave spirits into one bucket. Some focus on spirits served straight up, while others display their craft by mixing marvelous cocktails. Some bartenders—through no fault of their own—are rarely asked by their clients for anything other than their favorite pint. But there are also mixologists who refuse to offer any beer at all, and do not serve any "canned" or clichéd cocktails. The only agave spirits that some may serve is a "house tequila" paired with a lime, some salt, and a joke about worms or blacking out.

Nevertheless, more and more bartenders know how to access the best *mezcales* in the world, how to spin the richest stories about them, and how to serve them in cocktails that enhance their qualities. When serving a flight of mezcals for sipping, they offer them in handmade ceramic cups or cus-tomized *jicaritas* from the area of origin. They walk their patrons through the provenance information on the label, from distiller and locality to agave species and style of *alambique.*

Just as many brands now do, Siembra prints detailed, batch- and bottle-specific production data on every label, but those crucial details are only points of reference for a much larger story that is best told orally, face to face, *jicarita* next to *jicarita.* There is no label design that can convey all the environmental, historical, and social complexity of the agave spirit within.

The essential conduit to unpack that complexity is no less than the informed and engaged bartender, who can interpret specific points of pro-duction within the scope of the larger story and convey the narrative to the ear as well as palate. But what they are really trying to do is give each drinker a sense of the distinctiveness—the heartfulness—of every mezcal.

Bartenders are in the trenches, zipping around but with an aplomb that builds connections among consumers, producers, importers, distributors, and, sometimes, foreign legislatures as well. At the end of the night, the bartenders are the high priests—poised at the epicenter of spirits knowledge—who occupy a position of trust.

Such trust can be leveraged every-which-way. Some bartenders are on the payroll of spirits conglomerates, paid to continue a marketing push until liquor hits lip, and even after. Others, however, find a way to resist the pressures and propaganda by making purchasing decisions at the bar, not at corporate headquarters.

A few, like Houston's Bobby Heugel, have had the courage to dump thousands of dollars of dubiously promoted liquor down the drain and post their act of resistance on Instagram. The likes of bar legends like the late Tomás Estes, Phil Ward, Ryan Fitzgerald, Jeff Morgenthaler, Jim Meehan, Misty Kalkofen, Ivy Mix, Don Lee, Doug Smith, and Brian Eichhorst have earned their reputations by creating awareness and flexing their muscles. They have become much like well-regarded chefs who pioneered sustainability, food justice, and terroir in the Slow Food movement.

But when will the Slow Spirits movement gain equal footing with the Slow Food movement? (After all, our agaves and sotols grow far more slowly than Granny Smith apples, Gilfeather turnips, and geoducks.) Very few "slow bartenders" have yet to purge their restaurants of cheap *mixto* margaritas as Bobby Huegel has done in his bar. When will we begin discussing the irony of so-called farm-to-table restaurants that so carefully select their ingredients and produce but still serve tequilas produced with the harmful pesticides? We look forward to the days when "sustainable" restaurants have sustainable backbars.

For at least three decades, the executives driving the biggest brands in the tequila industry have been stripping away the deeply cultural underpinnings of Mesoamerican and Arid American beverages, opting for marketing their brands on their superficial associations with celebrities or beauty queens. They have done away with the use of fully seasoned, mature agaves and spring waters with superb minerality. Many have opted for products made with diffusers, autoclaves, and artificial flavors and colors, not because they are sadistic, but because it is profitable. They have even tried—unsuccessfully—to prevent small producers from using the word *mezcal*.

Our goal is to set the bar of integrity and quality higher. Once Siembra Spirits and the Tequila Interchange Project began to bring bartenders in to see what tequila had become, and what could yet be recovered, those bartenders began to refine their messages to their favorite customers back home. But the big players in the spirits industry soon countered, desperately trying to control the narrative. We witnessed a marked increase in the number of big brands that added bartenders—not just distributors—to their payroll as "ambassadors."

A few bartenders have become global ambassadors or sommeliers for hire by conglomerates with dubiously distinctive products, for they have been unable to resist a handsome paycheck. In exchange for their check, they must drink and promote *mixto* tequilas. They begin to forget self-evident truths and become blind to the shortcuts and bastardizations evident in broad daylight at many of the most well-endowed tequila distilleries.

Of course, there is plenty of information *and* misinformation about agave spirits available online, but what litmus test of veracity has been codified? Many novices are simply overwhelmed, if not baffled by it all. Bartenders, not brands, can teach customers enough about the trends in agave spirits that that they can use their own critical thinking to discern truth from fiction, and authenticity from fluff. That is, *if* bartenders and *cantineros* begin to better position themselves as the independent communicators-in-chief for the entire beverage supply chain. As soothsayers and truth testers, they hold the keys to the future of agave spirits.

As we witnessed with the sausage-making machinations that went into the NOM 186 for *mezcal*—which, fortunately, failed—the voice of bartenders still holds currency. They are the cross-pollinators that keep the many players and products of the spirits industry from becoming too inbred.

Both of us have been riveted by what happened in 2015 at a landmark meeting of forty bartenders and agave spirit producers on the patio of Bar Agricole in San Francisco, when they waged their first successful "intervention." Ryan Fitzgerald, co-owner of the Mission District's ABV hosted the meeting to reveal the sorry state of mezcal regulatory policies.

After thanking everyone for taking time out of their work schedule to participate in the strategy session, Ryan reminded them of what was at stake. It was the integrity of mezcals, and the capacity of traditional *mezcaleros* to adhere to their time-tried local traditions of craftsmanship under the pressure of multinational enterprises that were pushing mezcal regulations in the direction of tequilas and whiskeys.

As Bobby Heugel of the Anvil Bar and Refuge lamented that day, "The difficult part of this is accepting that new innovations in industrial production are going to be a part of mezcal. As a whole, industrial production threatens mezcal and may collapse the entire system."

This all brings us back to how bartenders directly influence the individuals or "consumers" who ultimately pay for the agave spirits. If consumers are hell-bent on purchasing a spirit because of an ad or an article that piqued their interest, they will endeavor to find it wherever it may be available, even if there is no rational reason for favoring that spirit over another. The bartender can be a counterforce, simply asking the customer to "taste and see" before jumping to conclusions that a highly touted spirit is the be-all and end-all among agave distillates.

Given this immutable fact of human behavior, we are left with two clear paths for shifting the trajectory of the spirits supply chain. We can either aid consumers who are open to shifting their preferences based on more knowledge and experience, or we make certain questionable spirits unavailable in our own programs.

Keep in mind that well into the last century, some poorly made mezcals were taken off the market. Some simply fell by the wayside due to lack of consumer confidence in the ways that they were made. In the northwest of Mexico, bootlegged distillates called *mezcal corrientes*—ones with nicknames like *chicote, chinguritos, margayate,* and *tumbayaquis*—went out of fashion because of their high methanol content, crude *venencia*, and poor selection of (often immature) agave plants. Inevitably, some crummy mezcals and tequilas ultimately fall by the wayside.

When poor-quality products are foisted upon them merely to reduce operation costs, bartenders can play a key role in saying *"Basta!"*—enough is enough. About the latter, we think of Max Reis of the California bar Gracias

Madre. After a trip to Jalisco and Michoacán, Max adamantly and very publicly eliminated all diffuser-made tequilas from his bar.

We reflect on Bobby Heugel, the most visible leader in Houston's hospi-
tality industry, who began a social media trend of dumping cases of Flor de
Caña rum down the drain when its Nicaraguan corporation was so severely
restricting water access to its farmworkers that some were dying of kidney
failure. (As a result, Flor de Caña came under such a firestorm of scrutiny
that the enterprise made vast and admirable improvements to the potable
water infrastructure in their fields.)

These are dramatic stories, but they are stories of eventual progress. Of
course, some companies need constant nudging to really get things right,
not just put up a smokescreen. California's Gracias Madre restaurant fur-
thered an industry trend of questioning and rejecting diffuser tequilas, to
the point where companies who formerly took pride in their efficient operation went to great lengths to hide their use of the machines.

Most bartenders do not make purchasing decisions, and they are certainly not at liberty to dump thousands of dollars of liquor down the drain
and post it on Instagram. But they are at liberty to speak their minds, and
we are all best served when those most at liberty to speak are exceptionally
well informed. Bartenders can use their unique power to educate and create
awareness of the trends of consumption, and thus trends of purchase.

Jeffrey Morganthaler explains his own teaching philosophy in this manner: "With my staff, I've always done small classes on different spirits
[where] I really push the idea, the fact, that *tequila is an agricultural product*,
and really try to talk about that difference between agricultural products
and industrial products and try to drive home that that's why this is different. . . . It does not really matter if a vodka producer is super responsible or
not. It doesn't have a huge impact, because it's an industrial product. But
with mezcal, it does."

Our ultimate goal is not to see that big companies are brought to their knees,
just that they be held accountable. To that end, we can leverage shifting purchasing trends, powered by people who care and communicate, so they

realize that a socially and ecologically unsustainable path forward will be equally unsustainable for their near-term bottom line. Then, and only then, can we hope for change.

Ryan Fitzgerald is one of those in the agave spirit world who explicitly articulated his "theory" for generating positive change: "When smaller, culturally sensitive brands openly communicate where their agaves come from, how old the agaves were at harvest, who was involved in the process from harvest to distillation, and how they execute each step of that process, the hope is that they put pressure on larger brands to do the same."

"At the very least, this transparency communicates to customers and industry decision makers which brands *are* worth more per bottle and which brands to support. Additionally, this transparency supplies consumers and the industry with tougher questions to ask of other brands—whether new or established, large or small—and those questions apply some pressure to those brands and their representatives to learn how to answer them."

In April 2020, we spent a Sunday evening with six of the young bartenders who were reinvigorating the spirits scene in Guadalajara's most innovative bars and restaurants. We had already bumped into them over the previous two weeks at Pare de Sufrir, De la O, Alcalde, La Tequila, Café palReal, and emerging hipster haunts too numerous to name. The average age of the *cantineros* present that night was easily half our own ages; they were men and women, native Tapatios or nomads of Mexican gastronomy.

As we shared sips of three mezcals and some small dishes with them, we asked them to speak of the challenges that they faced. They opened their hearts to us. We listened as intently as we could because it was clear that their generation was changing the way we think about and drink agave spirits.

Some of them talked about "putting in time" to learn their craft where a bar owner or senior bartender controlled both the list of spirits that could come in the door and the cocktails permitted to be created from them. Yes, they earned their chops under such mentors, but they longed for more.

So, they began to visit *maestros mezcaleros* on their days off, to taste and see the astonishing diversity of spirits emerging in Mexico. And that's when their own "conversion" began—one that allowed them to see agave sprits in the deeper way that Max Reis has described so well: "You realize this spirit is incredible because it's this picture of this time, and this place, and this terroir, and this family, and this tradition. . . . So, you start to see the way things were and how they can be and then you start to see the way things are when you juxtapose it against other practices. . . . I think it becomes very clear why you should be having certain practices and why these things should be preserved."

That is the moment when bartenders begin to respond to each mezcal for its distinctive character, and see how they can showcase it. It inspired the Tapatio bartenders to experiment with planning different mezcals in cocktails with *tepaches, ponches,* and locally fermented beverages, cured fruits, or fresh spices. They learned the vernacular language of the craft distillers for the flavor and fragrance profiles in order to enhance them rather than bury them under a ton of sugar. This became their way of showing respect for the efforts made by the *mezcaleros* themselves.

Back at the bar, the newly inspired *cantineros* of Guadalajara were at first disappointed by how few guests—Mexican citizen or traveling stranger—expressed initial interest in broadening their horizons. But gradually, step by step, drink by drink, their most allegiant customers became increasingly curious. As one bartender noted, "Sometimes we have really wonderful guests who come in and they want us to talk about our most unique products on the backbar."

The stories that these *cantineros* brought back from the field began to brighten faces and open new possibilities. That said, they conceded that they were all still involved in confidence-building with their most inquisitive clients. They could not beat them over the head with information about autoclaves, diffusers, nectar-feeding bats, or regulatory policies. But they could open conversations and gauge their customers' interest.

Their comments reminded us of what Max Reis said so well in one of our earlier encounters: "If you walk into a bar, whether it be a wine bar, or an agave bar, or a whisky bar, and you're with people that you value their opinion of you, the last thing you want to do is be shamed by the person that's taking care of you. So, we don't say anything negative about brands, we only build up the brands we're currently building up—and we also build up our guests,."

Nonetheless, all the *cantineros* we met in Guadalajara expressed heartbreak when David revealed to them just how little those at the front of the value chain—the farmers and fermenters, harvesters and haulers—receive in exchange for their sweat and blood, art and intelligence. And yet, as one teary-eyed women mourned, how do you start a more serious conversation about such things in places that people come for fun, a sanctuary where debates about politics and social justice are often kept at bay?

All were deeply moved by the many dimensions of our evening conversation, as it delved into principles, policy, pleasure, and promise. One young man stayed quiet for most of the night, but then asked us, simply, what we thought *he* could do—as one weary, overworked bartender—to make the industry healthier, saner, or more sustainable. How can he fit more hours into his day to explore such heady questions?

There is no easy answer, only gratitude and solidarity.

As David and I drove back to our rooms for the night, we were both humbled and inspired by the talent and commitment expressed among those young innovators. Although they are still daily dealing with many dilemmas that we cannot help them solve in our own limited lifetimes, we could rest assured that the next generation was on fire about the future of mezcal and the destiny of all other agave spirits.

They, not us, would lead the way.

Dr. Gentry, pioneering agave botanist, embraced by a maguey

Epilogue

Policies for a Brighter, More Delicious Future

Where do mezcals and other agave distillates go from here? If their distinctive identities and flavor profiles are to survive and further diversify to enhance our enjoyment, agave spirits must now and then spiral back to their ecological and cultural roots as the plants themselves do. As our friend Patricia Colunga-GarcíaMarín has said many times in many ways, "Our future is ancestral."

But the agave spirit economy must also spiral back to fully support agave nursery workers, farmers, harvesters, and small-batch distillers. Those engaged in key interactions in the production of agave spirits must ensure that bat pollinators, fuelwood trees, soil microbes, and fermentation vat microbes are taken care of, for they are on the front line. How do we as consumers listen sufficiently to agave farmers and their workers so that they know we support their efforts to make their operations saner and safer? And how do we form bridges between producers and consumers so that we all share in a holistic and whole-hearted vision of a healthier, more delectable future, one that we can savor for many more decades?

In Appendix 1, we offer a ten-point Mezcal Manifesto that establishes just such a holistic agenda for agave spirits—an agenda rooted in ethical values and a deep love for agaves themselves.

The deep roots that can best anchor the future of the agave family of distillates were most clearly expressed by one of our mentors in a vision that came to him in a dream over fifty years ago. Let us tell you the story of that remarkable vision and how it began to take root in the world.

※

On a hot but breezy summer evening in Tucson in 1978, a spry seventy-five-year-old plant explorer named Howard Scott Gentry joined two dozen of us in a desert garden to share *tragitos* of *mezcal bacanora* and to tell us stories. Dr. Gentry was at least twice the age of nearly all of us who attended that gathering in his honor, but he remained just as fit, witty, and "with it" as anyone in this wild bunch. After strolling around with a couple of his oldest friends in Tucson, identifying all the agave species in the garden, he was ushered to a lighted podium in the garden, where he pulled a handwritten manuscript out of his pocket. He took one more sip of *bacanora*, then put his cup down next to several pages of paper on the podium. He began to speak to us as a desert prophet or Mesoamerican shaman might do to their followers.

We were spell-bound by the very first words that came out of his mustached mouth.

"Being a man, I think and speak as a man, but tonight I also speak for agave. As agave's advocate, I speak in the same spirit as the bumper sticker that asks, 'Have you thanked a green plant today?' My aim is to stress the theme of mutualism among two disparate organisms. At the cost of being somewhat anthropocentric, I mean to show you as much what agave did for man as what man did for agave. The main pageant of this symbiosis took place in México, all the way back in time from the early hunting and gathering stages, through agriculture, and on into the city-state stages of civilization."

The warm light of the full moon offered a radiance to the old man's face. His voice, cleared and made crisp by his sips of a *mezcal joven*, made it seem as if he were declaring truths not just to those of us who were physically present that night, but to all the women and men who had ever imbibed the juices of agaves—fresh, fermented, or distilled—across all the ages.

Other than the plaintive calls of a poorwill and the hoots of a pygmy owl, you could hear a pin drop whenever Gentry occasionally paused to take another sip of mezcal. To a person, we were mesmerized. It was as if we had slipped into a collective trance.

Gentry veered off script now and then as elders often do, once to express his gratitude to his field companions in the barrancas of the Sierra Madre Occidental and the llanos of the altiplano: folks like Juan Arguelles, who guided Gentry on mules into Guarijio Indian country while he was still a teenage boy called Juanito; and Efraím Hernández Xolocotzi, the Mexican plant explorer and ethnobotanist Gentry admired above all others. But sooner or later, he returned to the point of his talk—how we as dwellers of the Americas are as dependent upon agaves as they are dependent on humans through our diligent care for them and conservation of their diversity:

"The ancient Mexicans cultivated and coddled the maguey incessantly. They cleared the wild land and put agave into it. They opened a new and nurturing environment with varying habitats and ecological niches for the random variants of the gene-rich agave genus. The cultivators made agave a home on the deep productive soils and in time provided water and manure. They protected the plants from weedy trees. They selected the genetic deviants of high production by planting genetic offsets. Agave species multiplied into more varieties than man has been able to characterize and count. This agricultural effort formed a socially disciplined complex subservient to the agave symbiont. Generally, that is what man did for agave in this Mesoamerican symbiosis."

Dr. Gentry savored another sip of mezcal to wet his whistle, then went on.

"In turn, agave has nurtured man. During the several thousand years that man and agave have lived together, agave has been a renewable resource for food, drink, and artifact. As man settled into communities, agaves became fences marking territories, protecting crops, providing security, and ornamenting the home. Agave fostered in man the settled habit, the attention to cultivation, and the steadfast purpose through years and life spans, all virtues required by civilization. As civilization and religion developed, the nurturing agave became a symbol, until with its stimulating juice man made a god out of it. Agave civilized man. That is what agave did for man!"

We could see our colleagues nodding, sighing, almost swooning, as Gentry expanded his vision out further and further, until it seemed to ripple out to every imaginable interaction between humans and nature. And then, the elderly botanist brought us back to the center of the vortex again, the spirits of agave.

"The beverages of agave are of special interest because they affect the mind as well as the digestive tract. . . . Among many, [agave] alcohol is a palliative because it permits mental and psychic exercises not otherwise functional. It promotes fellowship and communication; many a bargain or arrangement is made today in the social office of the cantina or club. As a social catalyst, [agave spirits] appeared to have fostered mental gyrations. . . . Mayahuel, the principal goddess of maguey, became a surrogate that nourished the body, slaked the parched throat, relieved the duty pressures, exalted the spirit, and provided at least temporary surcease from the hard life, and being god-like, protected the home. Altogether, this was another set of contributions of agave to man during the centuries of symbiosis. We see agave hosting man in a kind of social eroticism."

Never, *never,* had most of us heard such an accomplished scientist make such poetic leaps and mystical bounds. But as Gentry downed the last bit of mezcal in his cup, he put up a photo that someone had taken of him in the field years before, a diminutive figure sitting deep within the embrace of the upward curling leaves of an enormous *maguey de pulque*: "This is the last slide, and it is time for the last truth. You see me held in the arms of a giant maguey. I am a son of Mayahuel, the goddess of agave. *What I have told you today is what she told me to tell you.*"

Stunned silence. Then a few giggles, a pounding applause, whoops, and hollers. People were jumping up and down with delight, wonder, and laughter. Frosted glasses of margaritas were held up high above our heads, and a bottle of *bacanora* was passed around among those who drank it unadulterated. A couple of women went up and hugged the wizened field man. He beamed with a big grin.

Dr. Gentry accepted all this gratitude graciously, then held up both of his hands to calm us down. He asked us to listen for just another moment, to a part of his vision that he had not set down in print with the rest of his words:

"I sense that many of you would appreciate one more part of this story that I left out of my transcript, because it has not yet happened. It is about the future, not about the past. You see, one night a few years ago, I had a dream, one that I can remember with uncanny precision to this very day. I am still giving it my protracted attention, because I sense it has some viability for our future and for that of the agave.

"After coming off a rigorous plant-collecting trip, I was thoroughly exhausted, so I had a quick nightcap of mezcal, then fell asleep in a reclining chair. And then it came to me: the urgent need to make a better, safer place for all agaves in the world, given their cultural if not spiritual significance as well as the degree of peril that many plant species are now facing.

"In this dream, I could see a possible solution: the construction of a vertical botanical garden or sanctuary for all species of agaves in the Americas. It would be situated on the cliffs above that magical pueblo of Tepoztlán in the state of Morelos. It would be nestled in just below the remains of the ancient El Tepoztéco temple, which is built on top of the sacred Tepoztéco Mountain that overlooks the entire valley."

Gentry was speaking of a 30-foot (10-meter) pyramid built in the post-Classic period to honor Tepoztécatl, the Aztec god of harvest, fertility, the maguey plant, and its intoxicating beverage, pulque. In Mesoamerican mythology, Tepoztécatl was also one of the divine rabbits or "drunken bunnies" who gathered frequently to imbibe agave drinks together, just as we were doing that very night. Then Gentry said soberly, in a somewhat professorial tone, "There on the cliff face, I have observed a variety of microenvironments that would suit almost the entire range of agave species from all over the New World—sunny and shaded niches, wet or dry, of several kinds of rock and soil types. It is where the pollinators and soil biota already reside. Elevators could be easily built up against the cliff faces so that botanists, horticulturists, and farmers could climb with ease between patches of agaves that grew best at different elevations; horizontal ramps, scaffolding, or moving sidewalks could safely transport their caretakers along the breadth of the mountainside.

"At this living gene bank, conservatory, or in situ reserve of agave diversity, conservationists could move pollen grains betwixt and between different agaves as they flower to hand-pollinate them for purity's sake, or let the many nectar-feeding bats frequenting the caves of Tepoztéco Mountain cross-pollinate any open flowers, thereby generating more diversity. This conservatory would not replace the nature reserves in the home range of each species but would serve as a safety net should anything happen to the rarest of those agave populations left in the wild. And if I live long enough, I hope to work with laddies and lassies like you all to bring this dream into fruition."

Although Gentry lived another decade before his death at the age of ninety, he never saw his epiphanous vision for a Tepotzlán agave conservatory materialize. He did, however, continue to work at his Gentry Experimental Farm in Murrieta, California, growing dozens of species of agave, and mentoring younger agave researchers at botanical gardens and universities in Arizona, California, and many states in Mexico. These "young sprouts" of human-kind—like the vegetative offshoots of a maguey or the divine rabbits—are also the progeny of the agave goddess, Mayahuel.

You too are now part of the grand agave family as well. With your help in your roles as the Centzon Tōtōchtin—the Aztec spirits of party-going, agave-drinking children of Mayahuel—we just may bring Gentry's vision to fruition before we all leave Planet Desert for the Agave Spirit World.

If you have made it this far through this story of the wonders and woes of those mandala-like succulents, you too have become part of that ancient symbiosis between man and maguey, between woman and Mesoamerican mezcal. We appoint you as caretakers of that dynamic symbiosis and urge you to ensure that these spirited relationships are never extinguished.

For all the love we (and you) may have for the many mezcals and the creative innovators who make them, we are painfully aware that the agave spirits industry is facing daunting challenges throughout Mexico.

Wild agave foragers and farmers, firewood cutters and foresters, master distillers and bartenders, restauranteurs and distributors, consumers and conservationists must soon reckon with these challenges. Unless more sustainable practices and compassionate policies are put in place immediately, we will risk losing much of what is of cultural and ecological value in this realm.

We risk seeing more wild populations and even species of agave on the verge of extinction. We risk seeing once-common fuelwood species depleted. We risk losing the soil fertility, microbial diversity, and precious pollinators of agave-dominated landscapes. We risk witnessing the demise of already rare cultivated varieties of mezcal, and the concomitant impoverishment of traditional ecological knowledge about them. We risk wiping out the myriad yeasts and bacteria in the fermentation vats, and the knowledge of how to bring out their flavors and fragrances through careful distillation.

If the former abundance of these natural and cultural treasures declines further, we may also see the loss of the health, social well-being, and livelihoods of tens of thousands of workers in agave country. Consumers stand to lose the distinctive aromas and taste profiles of the most complex spirits on Earth—the many unique agave distillates that now delight so many of us. A ripple effect will emanate from such losses that will lead to the disruption of beneficial interactions among species, cultures, and nations.

Fortunately, the Mexico Secretariat for the Economy welcomes any stakeholder—harvester, distiller, distributor, bartender, or consumer—who cares enough to request a place at the table when the regulatory commissions for tequila, mezcal, and other spirits host decision-making meetings. With the proposal of NOM 186—aimed at restricting any use of the word *mezcal* by producers or distributors not under the umbrella of the Consejo Regular de Mezcal—thousands of Mexican and American citizens expressed their outrage. In this case, the opinions of mezcaleros, bartenders, and consumers held sway.

By the time NOM 199 was proposed to force producers of 100 percent agave distillates to use the name *komil* for their spirits and ban them from using the scientific term *agave*, more than twenty thousand signatures and letters stormed in to oppose this insult. As *Mezcalistas* has reported, "In a stunning reversal of fifty years of policy, the administration of President López Obrador has decreed that the nation of Mexico is giving the word *mezcal* back to the people of Mexico so that they can 'reinvigorate the Mexican soul.'"

The good news is that this surge in interest need not start from scratch in forging solutions to prevent what UNAM ecologist Alfonso Valiente-Banuet has called the "impending ecological and cultural collapse of agave landscapes." He has already called for "a comprehensive plan that combines the production of artisanal mezcal, rural development for small producers, and maintenance of biodiversity."

Valiente-Banuet is one of a dozen or more world-class ecologists and conservationists who have already worked out many of the elements to be integrated into such a comprehensive plan. To be sure, the plan cannot simply stop at preventing agave populations and species from going extinct. Valiente-Banuet and his colleagues have clearly warned that "a missed component of biodiversity loss that often accompanies or even precedes species disappearance is the extinction of ecological [and cultural] interactions."

So, true solutions must go beyond just halting the ongoing losses of one agave species after another. They must deal with what bat biologist Ted Fleming and Gary called "the conservation of New World mutualisms" in an op-ed in *Conservation Biology* in 1993. To avoid the extinction of relationships, long-term solutions must conserve and restore all kinds of species interacting in "the agave holobiont"—the microbes on agave roots and leaves, the pollinators and other beneficial fauna, the companion plants in *milpa* polycultures, and the community of human stewards.

Proven solutions are at hand. In fact, they are already being piloted on the ground in at least a half dozen Mexican states, so we do not wish to imply that these efforts are "starting from scratch." What we do propose is working collaboratively with many economic sectors and with Indigenous cultures to scale up efforts that put agave production and consumption on a more resilient trajectory.

Here are quick assessments of the eleven key problems, matched with our proposed solutions:

PROBLEM 1. The unbridled "boom" in demand risks the extinction of wild agave species. Already, two in every three agave species are facing some level of risk due to multiple causes, the demand for mezcal being primary for most.

    Solution 1: *Protect and restore the biological base for all agave distillates— the agave species, as well as nurse trees, soil microbes, and pollinators they rely upon.*

PROBLEM 2. Regulatory restrictions in tequila, mezcal, *bacanora,* and *raicilla* favor selected clones or subspecies over heterogeneity, risking genetic collapse, insect infestations, climate catastrophes, insect infestations, and disease outbreaks.

    Solution 2: *Promote a broader genetic base for production, including blends of multiple species or varieties.*

PROBLEM 3. Agave monoculture in *tequila, espadín,* and possibly *bacanora*—with a few clones or vegetatively propagated stocks planted over tens of thousands of hectares—risks agroecological collapse, disease pandemics, and insect plagues.

    Solution 3: *Reintegrate multiple species and varieties of agaves of different ages into diverse milpas and other agroforestry systems.*

PROBLEM 4. Same-age stands of agaves risks demographic collapse and exacerbates harmful fluctuations in the availability of agave nursery stock for outplanting.

    Solution 4: *Better employ the Sembrando Vida program, private growers, and other strategies to start up nurseries of more agaves grown from seed to foster a heterogeneity of age classes in fields so that 10 percent of plants fully mature to be harvested each year.*

PROBLEM 5. The harvesting of four- to seven-year-old immature agave plants to place into diffusers where their fructans can be heat-blasted into a sugar slurry is ecologically damaging and devoid of the flavors, colors, and fragrances that mature agaves impart.

Solution 5: *Ban the harvesting of immature plants, diffusers, and artificial flavorants and colorants from the elaboration of tequilas and other agave spirits.*

PROBLEM 6. Landscape uniformity and fuelwood depletion of trees disrupts ecological interactions, new plant recruitment, and establishment, pushing the ecosystem toward collapse.

Solution 6: *Safeguard older trees, agaves, and bat habitats (including caves and rock crevices) adjacent to fields and protect 30 percent of all reforested agaves from any kind of harvesting until after they flower and set seed.*

PROBLEM 7. Highly skilled and knowledgeable *jimadores* and *mezcaleros* with decades of agave stewardship are being replaced by unskilled day workers, risking the loss of agroecological knowledge for plant management.

Solution 7: *Provide better wages and benefits as well as other incentives to keep families engaged in agave stewardship.*

PROBLEM 8. Excessive use of pesticides and herbicides is damaging human health and risking collapse of the workforce, triggering cancer and early mortality among workers.

Solution 8: *Fully adopt integrated strategies for pest and disease management in agave fields to reduce the negative impacts of excessive agrichemical toxins, while providing better occupational health care and toxicological testing in local health clinics.*

PROBLEM 9. The transformation of hypoglycemic agave inulins into high-fructose syrups or "nectars" risks aggravating Mexico's worst public health crises—childhood obesity and adult-onset diabetes.

Solution 9: *Ban the use of agaves in diffusers and autoclaves that render high-fructose syrups and nectars, and place warnings on all labels of agave sweeteners to alert consumers of the health risks of excessive consumption of simple sugars.*

PROBLEM 10. The monopolization of markets and long-distance transport of agave planting stock from one production area to another risks price gouging, the spread of crop diseases, and the homogenization of terroir, and negates the integrity of place-based appellation of origin.

Solution 10: *Strengthen phytosanitary protocols, restrict or ban extra-local transfer of planting stock, subsidize community-owned nurseries, and place price ceilings on agave planting stock.*

PROBLEM 11. The current disparities in compensation for work in the agave supply chain are among the worst in the food and beverage industry worldwide, favoring multinational corporate executives, planting stock intermediaries, and retailers at the expense of harvesters, distillers, and bartenders.

Solution 11: *Restructure the agave spirit supply chain to eliminate any kind of subsidy going to larger corporations and create more equity among the players in every link within the supply chain, so that both risks and benefits are more evenly spread.*

For such solutions to work, distillers may have to cut their ties with regulatory councils and use alternative sources for lab testing, export promotion, collective marketing, and policy setting. More producers will opt out of calling their products anything but 100 percent agave distillates but should consider collective trademarks and geographic indicators or appellations instead of dancing to the Denomination of Origin drum.

In short, there will be strong resistance from producers, bartenders, and consumers to any other attempts to control or homogenize agave spirits. New, more equitable structures such as civil associations and farmer-distiller cooperatives will likely emerge with greater frequency in every state where agave spirits are elaborated, rather than the industry proceeding farther down the precarious path of the big tequila industry.

When these other trajectories generate healthier, better-compensated workers, and more diverse and flavorful spirits without further depleting nature or culture, the responsible drinkers of the world will stand up and take notice.

That will be the moment when all of us can lift our *jicaritas* high to offer a blessing, prayer, or cheer, and drink mezcals and their many spirited kin without any regret.

Acknowledgments

David and Gary wish to acknowledge the good will and encouragement of our wives—Marité and Laurie, respectively—and our coworkers while we were immersed in the travel for the project, and during the writing of multiple drafts. We are grateful for the involvement of our kids—David Jr., Elisa, and Dan Marcos; and Laura Rose, Danny, Jeremy, Jessica, and Dustin—during many agave-related adventures and events. Our professional team members Oscar Serrano, Dino Rosario, Eduardo Moreno, Maria Cisneros, Nicole Harris, Ana Urgiles, Brenda Padilla, Ana Padilla, Ricardo Cardenas, Jeff Banister, Ben Wilder, Erin Riordan, and Beto Villa offered us extraordinary help and guidance.

We are also indebted to our confidantes and mentors who have guided us for decades, including Tomás Estes, Howard Gentry, "Xolo" Hernandez X., Carlos Camarena, Salvador Rosales Briseno, Salvador Rosales Torres, Patricia Colunga García Marín, Daniel Zizumbo Villareal, Pedro Jiménez Gurria, Alfonso Valiente Banuet, Arturo Gomez-Pompa, Suzanne and Paul Fish, Alan Weisman, William Steen, Paul Mirocha, Ana Guadalupe Valenzuela Zapata, Exequiel Ezcurra, Luis Eguiarte, Valeria Souza, William "Doug" Smith, Rodrigo Medellín, Rogelio Luna, Rodolfo Fernandez, Claudio Jimenez, Fernando Gonzalez, Jose Hernandez, Wendy Hodgson, Greg Starr, Alejandro de Ávila, Alejandro Casas, Francisco Terrazas, Monique Huston, Juan Olmedo, the late Ronnie Cummins, and Alfredo Corchado.

We wish to express our gratitude to the many families of mezcaleros with whom we have interacted over the years: las familias Vieyra, Rosales, Ángeles Carreño, Olivera Aguilar, Contreras, Partida, Fernáandez, Macedas, Sánchez, Encinas Molina, Miranda Parada, Juarez, Virgen, Perez, Joya, Romero, among many others. We are grateful that agave conservation and culture are parts of agave festivals in Tucson, and Marfa, and the Fermentation Fest in

Wisconsin. Special thanks to Todd Hanley and Francisco Terrazas of Hotel Congress, Jonathan Mabry of Tucson City of Gastronomy, and Tim Johnson of Marfa Book Company, Jay Salinas and Donna Neuwirth, and to Meredith Dreiss, Elizabeth Johnson, Sarah Bowen, and Sarita Gayatan, who enrich these events. David wishes to thank the following bartenders, mixologists, and restaurateurs: Federico Diaz de Leon, Phil Ward, Misty Kalkofen, Bobby Heugel, Ryan Fitzgerald, Jim Meehan, Joaquin Meza, Jef Morgenthaler, Maxwell Reis, Tetsu Shady, and Ivy Mix.

Gary and David wish to thank the following academics, nursery keepers, and bat and agave activists: Cesar Ojeda-Linares, Patricia Carlos Martinez del Río, Ted Fleming, Park Nobel, Abisai Garcia Mendoza, Donna Howell, Robert Bye, Jr., Patricia Lappe-Olivera, Nacho Torres, Raul Puente, Luis Hernandez, Ryan Stewart, Greg Starr, the late Tony Burgess, Edmundo Garcia Moya, and Iván Saldaña Oyarzábal.

We are grateful to those involved in the editing, art, and production of this book: René Tapia, Melanie Tortoroli, Annabel Brazaitis, Jessica Murphy, Karen Wise, Sean Duffin, and Anna Oler.

Gary is grateful for support for the W. K. Kellogg Endowment at the University of Arizona, Agnese Haury, Desert Botanical Garden, Arizona-Sonora Desert Museum, and Borderlands Restoration Network.

Our entire team is grateful to literary agents Victoria Shomaker and Richard Friedman for helping to conceptualize this book and for facilitating it reaching the right hands.

Appendix 1

The Mezcal Manifesto

Gary Paul Nabhan and David Suro Piñera

Note: This is a "call for action" that extends upon the short list of problems and solutions at the end of the last chapter. It draws upon the wisdom of many farmers, wild plant harvesters, distillers, bartenders, pollination ecologists, food justice scholars, microbiologists, and policymakers offered through dialogues between 2019 and the end of 2022. We hope to get it into the hands of Regulatory Commission representatives and federal officials during 2023.

Preamble

As harvesters, habitat restorationists, growers, guardians, distillers, distributors, botanists, bartenders, consumers, and concerned citizens whose landscapes and livelihoods depend upon the conservation of the agave plant, its cultural impact, and the culinary enjoyment of drinks and foods derived from agaves,

We join together to alert the world to the daunting challenges now facing at least 60 percent of all agave species, their associated wildlife, and microbial diversity, and the agave spirit industry that relies upon them.

As Mexican ecologist Alfonso Valiente-Banuet so clearly warned, "A

missed component of biodiversity loss that often accompanies or even precedes species disappearance is the extinction of ecological [and cultural]
interactions." Because of the ongoing losses not just of one agave species
after another, but of all the other species interacting in "the agave holobiont," we need what Dr. Valiente-Banuet refers to as "a comprehensive plan
that combines the production of artisanal mezcal, rural development for
small producers, and maintenance of biodiversity."

Together we seek not to critique, but to craft long-term solutions for the
equitable and sustainable production of the many mezcals and kindred
agave spirits.

We are dedicated to improving the well-being of all those involved in
every link in the agave spirit supply chain from botanists and cultivators to
bartenders.

We have arrived at a critical moment in history for determining the future
of agave distillates, which exemplify the unique contributions of Mexico to
the gastronomic patrimony of the Americas.

It is time to act.

Unique Values

We wish to honor, protect, and celebrate the distinctiveness of mezcals and
other 100 percent agave distillates relative to all other liquors produced on
the planet, including tequila:

1. There is more "biodiversity in a bottle" embedded in small-batch 100
   percent agave distillates than in any other alcoholic beverage in the
   world, with more than sixty-two species of agaves being used in mezcal
   production, in addition to more yeasts and bacteria than those involved
   in the fermentation of any other commercially available spirit.
2. The probable pre-colonial microdistillation of agave alcohol likely rep-
   resents the oldest distillation tradition for any plant-based beverage in
   the Americas and should be considered a key feature of the Mexican
   gastronomic heritage decreed by UNESCO as a Patrimonio Cultural de
   Humanidad, or a Cultural Legacy of Global Significance to Humanity.

3. More Indigenous and immigrant cultures have contributed to the traditional knowledge and biotechnologies of appropriate scale used in producing agave spirits than to any other beverage. These stakeholders live alongside and work with agaves in at least twenty-two Mexican states.

4. Many non-timber forest products are associated with mezcals as ingredients in agave-based herbal medicines, infusions *(curados)*, *pechugas*, and salts.

5. There are well-documented spiritual, ceremonial, and ritual uses of mezcals in many Indigenous cultures of Mesoamerica that should be protected as part of their inalienable rights to religious freedom of expression. Their intellectual property rights also need protection.

Ten-Point Action Plan

We propose a comprehensive action plan to set the agave spirit industry and its many stakeholders on a healthier, more sustainable course. With appropriate public investment, all of these proposed actions can be readily accomplished within a decade.

1. Escalate efforts in both Sembrando Vida and private initiatives to reforest both diverse wild agaves and trees needed as fuelwood in mezcal roasting and distillation. Three out of every ten of these agaves and trees should be left to flower for bats and other pollinators so that they can produce seed for forest regeneration.

2. Assure greater protection and withdraw permits for harvesting mezcal in any area in which wild agave species occur that are on the IUCN Red List of Threatened Species. Increase monitoring and on-ground protection of the eighteen agave species protected under the official Mexican Norm NOM-O59-SEMARNAT-2010.

3. Create a buffer between wild populations of agaves and any new plantings of the *tequilana azul* cultivar on rented or purchased lands to slow the spread of both the *tristeza y muerte* pathogens and *picudo* pest infestations.

4. Grant carbon offsets or payments to ecosystem services only to *ejido* collectives or property owners who have small holdings of private lands

and who grow four or more species of agave, rather than offering them largely to proprietors of monocultural plantings of *tequilana azul, espadín,* or *henequén.*

5. Establish a collective trademark for each local or Indigenous tradition of agave spirits, or an appellation of origin, geographic indicator, or collective branding other than the Denominations of Origin for Tequila and Mezcal. It should be managed by an *association civil* nonprofit directed by producers themselves, as alternatives to external Consejo Regulador entities that do not favor small-batch producers.

6. Invest in community-based technological infrastructure for cost-efficient local processing of agave inulins—not for high-fructose agave syrups and nectars, but for probiotic, antidiabetic beverages and foods to deal with Mexico's costly epidemic of diabetes and obesity.

7. Expand rural health clinics in areas of agave production to test for toxins derived from pesticides, herbicides, and other agrichemicals, to monitor and treat the local population, and to deal with the increasing frequency of heat-related accidents, strokes, dehydration, and exhaustion associated with climate change.

8. Grant all skilled distillers *(mezcaleros),* harvesters (including *jimadores),* and daily wage workers *(jornaleros)* in agave fields all health care and retirement benefits available through IMSS.

9. Subsidize development of on-farm infrastructure to use trimmed agave leaves from fields and bagasse from distilleries as fermented silage for livestock feed, or as organic soil amendments.

10. Certify (at much lower economic costs and using less paper) any uncertified, small-scale distillers who wish to be certified. Many now must clandestinely produce and sell their bootleg agave spirits because of high "pay to play" entry costs. Waive or reduce the Value-Added Tax (IVA) and Special Tax on Products and Services (IEPS) for small-scale artisanal producers, as is done for traditional producers of Indigenous crafts and other handmade products. These taxes now remove 59 percent of the value of each bottle of agave distillate sold in Mexico from the income of the producer.

Appendix Two

Wild Agaves and Cultivated Landraces (Razas Criollas) Distilled into Agave Spirits

This is the first state-by-state comprehensive index that cross-lists common names in Spanish and Indigenous languages with accurate scientific names for agave species. One of the principal sources is Colunga-GarcíaMarín, Zizumbo-Villarreal, and Martínez Torres (2007). Websites of *Mezcal Reviews, Mezcal PhD, Mezcalistas,* and *Todo Mezcal* periodically update their listings on line; these were primarily noted in the summer of 2021.

State & Folk Names	Species	Source
CHIHUAHUA		
blanco or tosá	*Agave shrevei*	Bye et al. 1975; *Mezcal Reviews*
casero or gusime (cultivated)	*Agave angustifolia*	Bye et al. 1975
chahui	*Agave multifilifera*	Bye et al. 1975
gubuk	*Agave angustifolia*	Colunga-GarcíaMarín et al. 2007
gusime or del monte (wild)	*Agave angustifolia*	Bye et al. 1975

State & Folk Names	Species	Source
ki'mai	Agave applanata	Colunga-GarcíaMarín et al. 2007
ku'uri	Agave applanata	Colunga-GarcíaMarín et al. 2007
sa'puli	Agave bovicornuta	Colunga-GarcíaMarín et al. 2007
verde or ojcome	Agave wocomahi	Bye et al. 1975
COAHUILA-NUEVO LEÓN		
cenizo	Agave asperrima	González-Elizondo et al. 2009
DURANGO		
cenizo or ji'ja	Agave durangensis	Gonzaléz-Elizondo and Galván 1992; Mezcal Reviews
cenizo, lechuguilla verde, or raicilla de la sierra	Agave bovicornuta	Mezcal Reviews
chacaleño	Agave angustifolia	González-Elizondo et al. 2009
chico de la sierra or niño	Agave maximiliana	Gonzaléz-Elizondo and Galván 1992
cimarrón, de la barranca, mayero, mezcalillo, or negro	Agave inaequidens	González-Elizondo et al. 2009
de castilla or kokma'mai	Agave applanata	Gonzaléz-Elizondo and Galván 1992
de la sierra or ji'ja	Agave shrevei var. magna	González-Elizondo et al. 2009
espadín	Agave angustifolia	González-Elizondo et al. 2009
gubuk	Agave angustifolia	Colunga-GarcíaMarín et al. 2007
jija	Agave durangensis	Colunga-GarcíaMarín et al. 2007
lamparillo	Agave asperrima	González-Elizondo et al. 2009

State & Folk Names	Species	Source
masparillo	*Agave maximiliana*	*Mezcal Reviews*
sapulh	*Agave bovicornuta*	González-Elizondo et al. 2009
tepemete or gubuk	*Agave angustifolia*	González-Elizondo et al. 2009; Gonzaléz Elizondo and Galván 1992
tequilana azul	*Agave tequilana*	González-Elizondo et al. 2009
verde	*Agave salmiana*	*Mezcal Reviews*
GUANAJUATO		
maguey pulquero	*Agave salmiana*	*Mezcal Reviews*
GUERRERO		
ancho or bravo	*Agave cupreata*	Figueredo-Urbina et al. 2017
cacalotentli	*Agave angustiarum*	Colunga-GarcíaMarín et al. 2007
chino or papalote	*Agave cupreata*	Figueredo-Urbina et al. 2017
delgado	*Agave angustifolia*	*Todo Mezcal*
espadín	*Agave angustifolia*	*Mezcal Reviews*
papalote	*Agave cupreata*	*Mezcal Reviews; Todo Mezcal*
JALISCO		
azul (tequilana)	*Agave tequilana*	Valenzuela-Zapata 2011; Trejo et al. 2018
bermejo	*Agave sisalana*	Valenzuela-Zapata 2011
cenizo	*Agave bovicornuta*	*Mezcal Reviews*
chato	*Agave americana*	Trejo et al. 2018
espadilla	*Agave angustifolia*	Colunga-GarcíaMarín et al. 2007

State & Folk Names	Species	Source
gigante	*Agave valenciana*	González-Elizondo et al. 2009
ixtlero amararillo	*Agave rhodacantha*	Trejo et al. 2018
ixtlero verde	*Agave angustifolia*	Trejo et al. 2018
masparillo	*Agave maximiliana*	Gonzaléz Elizondo and Galván Villarreal 1992
masparillo	*Agave bovicornuta*	Gonzaléz Elizondo and Galván Villarreal 1992
moraleño	*Agave sisalana*	Valenzuela-Zapata 2011
pata de mula	*Agave tequilana*	Valenzuela-Zapata 2011
pencudo	*Agave angustifolia*	Valenzuela-Zapata 2011
relisero	*Agave valenciana*	*Todo Mezcal*
sigüín	*Agave tequilana*	Trejo et al 2018; Valenzuela-Zapata 2011
zopilote	*Agave tequilana*	Valenzuela-Zapata 2011
MICHOACÁN		
alto	*Agave inaequidens*	Figueredo et al. 2014; *Mezcal Reviews*
bruto or mezcal alto	*Agave inaequidens*	Figueredo et al. 2014
bruto or chapín	*Agave inaequidens*	Figueredo et al. 2014
cenizo or negro	*Agave inaequidens*	Figueredo et al. 2014
chico or alto	*Agave inaequidens*	Figueredo et al. 2014
chino, cupreata, or papalote	*Agave cupreata*	*Mezcal Reviews*
grande or alto	*Agave inaequidens*	Figueredo et al. 2014

State & Folk Names	Species	Source
hoja ancha y espina grande or alto	*Agave inaequidens*	Figueredo et al. 2014
hoja ancho or alto	*Agave inaequidens*	Figueredo et al. 2014
hoja angosta or alto	*Agave inaequidens*	Figueredo et al. 2014
hojas largas y espina chica or alto	*Agave inaequidens*	Figueredo et al. 2014
manso	*Agave hookeri*	Figueredo et al. 2014
manso de la sierra	*Agave americana*	Figueredo et al. 2014
NAYARIT		
ceniza or lechuguilla verde	*Agave bovicornuta*	*Mezcal Reviews*
OAXACA		
arroqueño	*Agave americana* var. *oaxacensis*	CONABIO 2016
barril, b.chico, b. gordo, or b. verde	*Agave karwinskii*	CONABIO 2016
barril gordo	*Agave rhodacantha*	CONABIO 2016
barril verde	*Agave macrocantha*	CONABIO 2016
batobpaz	*Agave potatorum*	Colunga-GarcíaMarín et al. 2007
bicuishe	*Agave karwinskii*	Colunga-GarcíaMarín et al. 2007; CONABIO 2016
biliaá	*Agave seemanniana*	Colunga-GarcíaMarín, et al. 2007; CONABIO 2016
blanco	*Agave americana* var. *oaxacensis*	*Todo Mezcal*
cachitún	*Agave karwinskii*	CONABIO 2016
canastruda, canastuda	*Agave americana* var. *americana*	Graciela Carreño Ángeles, pers. com.
candelillo	*Agave karwinskii*	CONABIO 2016

State & Folk Names	Species	Source
ceniza	*Agave americana* var. *oaxacensis*	CONABIO 2016
chamisa	*Agave karwinskii*	Colunga-GarcíaMarín, et al. 2007
chato	*Agave rhodocantha*	CONABIO 2016
cirial	*Agave karwinskii*	CONABIO 2016
coyote	*Agave americana*	*Mezcal Reviews*
coyote	*Agave karwinskii x A. potatorum*	*Mezcal Reviews*
cuishe	*Agave karwinskii*	CONABIO 2016
curandero	*Agave marmorata*	CONABIO 2016
de carne or de lisa	*Agave sisalana* or *A. americana x A. angustifolia*	Graciela Carreño Ángeles and Sósima Olivera Aguilar, pers. com.
de Castilla	*Agave americana*	CONABIO 2016
de cuela	*Agave marmorata*	*Mezcal PhD, Mezcalistas*
de lumbre	*Agave angustifolia*	*Mezcal PhD, Mezcalistas*
de pulque	*Agave atrovirens*	*Mezcal Reviews*
de pulque	*Agave americana* var. *americana*	CONABIO 2016
de rayo	*Agave americana* var. *oaxacensis*	CONABIO 2016
do ba daan	*Agave rhodocantha*	CONABIO 2016
do-be, dobzan, duabla	*Agave seemanniana*	CONABIO 2016
duende/coyote	*Agave lyobaa*	CONABIO 2016; *Mezcal Reviews*
espadilla	*Agave angustifolia* var. *rubescens* and/or var. *angustifolia*	CONABIO 2016
espadilla	*Agave rhodacantha*	CONABIO 2016

State & Folk Names	Species	Source
espadin	*Agave rhodacantha*	CONABIO 2016
espadin	*Agave angustifolia* var. *angustifolia*	CONABIO 2016
huiscole	*Agave marmorata*	CONABIO 2016
jabali, jabalín	*Agave convallis*	CONABIO 2016; *Mezcal Reviews*
largo	*Agave inaequidens*	*Todo Mezcal*
madre cuishe	*Agave karwinkskii*	CONABIO 2016
madre cuishe penca larga	*Agave rhodacantha*	CONABIO 2016
mano largo	*Agave karwinskii*	*Mezcal Reviews*
marteño	*Agave karwinskii*	CONABIO 2016; *Todo Mezcal*
Mexicano, m. amarillo	*Agave rhodacantha*	CONABIO 2016
pitzometl, pichomel	*Agave marmorata*	CONABIO 2016
papalometl	*Agave seemanniana*	CONABIO 2016
papalote, papalometl, mariposa	*Agave cupreata*	CONABIO 2016; Todo Mezcal
pulquero	*Agave americana* var. *oaxacensis*	CONABIO 2016
ruqueño	*Agave americana* var. *oaxacensis*	*Todo Mezcal*
serrano	*Agave americana* var. *americana*	*Todo Mezcal*
sierra negra	*Agave americana* var. *oaxacensis*	CONABIO 2016
sierrudo	*Agave americana* var. *americana*	CONABIO 2012; *Todo Mezcal*
tepeztate	*Agave marmorata*	CONABIO 2016
tobalá	*Agave cupreata*	CONABIO 2016
tobalá chato	*Agave seemanniana*	*Mezcal Reviews*

State & Folk Names	Species	Source
tobalá chino	*Agave potatorum*	*Todo Mezcal*
tobalá orejón	*Agave potatorum*	*Todo Mezcal*
tobasiche or tabaxiche	*Agave karwinskii*	CONABIO 2016; *Mezcal Reviews*
tripón	*Agave karwinskii*	CONABIO 2016; *Mezcal Reviews*
verde	*Agave karwinskii*	*Todo Mezcal*
xolo	*Agave americana* var. *oaxacensis*	CONABIO 2016
yabadensii	*Agave cupreata*	Colunga-GarcíaMarín et al. 2007
yavicuam	*Agave americana* var. *oaxacensis*	Colunga-GarcíaMarín et al. 2007
yavitcuishi, papalometl	*Agave nuusaviorum* subsp. *nuusaviorum*	CONABIO 2016; *Slow Food Mexico*
PUEBLA		
cachutum	*Agave karwinskii*	Colunga-GarcíaMarín et al. 2007
espadilla	*Agave americana*	*Mezcal Reviews*
papalometl	*Agave potatorum*	*Todo Mezcal*
pichomel or pizometl	*Agave marmorata*	*Todo Mezcal*
SAN LUIS POTOSI		
cenizo	*Agave americana* subsp. *protoamericana*	*Todo Mezcal*
Cimarrón, manzo, verde, or i'gok juguiarum	*Agave salmiana* subsp. *crassispina*	González-Elizondo et al. 2009
lamparillo	*Agave asperrima*	González-Elizondo et al. 2009
mexicano	*Agave americana*	González-Elizondo et al. 2009
lechuguilla	*Agave univittata* subsp. *lophantha*	González-Elizondo et al. 2009
salmiana or pulquro	*Agave salmiana*	*Mezcal Reviews*

State & Folk Names	Species	Source
SONORA		
a'ud nonhakam	*Agave murpheyi*	Nabhan 1985; Hodgson 2001
bacanora	*Agave angustifolia*	Gentry 1982
ceniza	*Agave colorata*	Gentry 1982
chino	*Agave rhodocantha*	Holguín 2020
jaiboli or temeshi	*Agave jaiboli*	Gentry 1982
lechuguilla del norte	*Agave palmeri*	Gentry 1982
lechuguilla del sur, blanco, or ceniza	*Agave shrevei*	Gentry 1982
lechugilla verde or sapari	*Agave bovicornuta*	Gentry 1982
noriba	*Agave bovicornuta*	Colunga-GarcíaMarín, et al. 2007
San Antoneña	*Agave rhodacantha*	Holguín 2020
tauta	*Agave parviflora*	Gentry 1982
wocomahi	*Agave wocomahi*	Gentry 1982
yocogihua or maguey verde	*Agave rhodacantha*	Gentry 1982, Holguín 2020
TAMAULIPAS		
amole	*Agave univittata*	*Mezcal Reviews*
áspero	*Agave scabra* ssp. *scabra* (= *A. asperrima*)	Jacques-Hernández et al. 2007
burgos	*Agave americana*	Jacques-Hernández et al. 2007
cenizo	*Agave scabra* ssp. *scabra* (=*A. asperrima*)	Jacques-Hernández et al. 2007
cruillas	*Agave americana*	Jacques-Hernández et al. 2007
espadilla	*Agave angustifolia* var. *angustifolia*	Jacques-Hernández et al. 007

State & Folk Names	Species	Source
jarcia	*Agave gentryi*	Jacques-Hernández et al. 2007
jarcia	*Agave montium-sancticaroli*	Jacques-Hernández et al. 2007
jarcia/montana	*Agave montana*	Jacques-Hernández et al. 2007
lechuguilla	*Agave funkiana*	*Todo Mezcal*
lechuguilla	*Agave univittata* subsp. *lophantha*	*Todo Mezcal*
mezortillo	*Agave univittata* subsp. *lophantha*	*Todo Mezcal*
San Carlos	*Agave americana*	Jacques-Hernández et al. 2007
San Nicolas	*Agave americana*	Jacques-Hernández et al. 2007
zapupe verde	*Agave angustifolia* var. *deweyana*	Jacques-Hernández et al. 2007
TLAXCALA		
pulquero	*Agave salmiana*	*Mezcal Reviews*
YUCATÁN		
chelem amarillo, blanco	*Agave angustifolia*	Colunga-GarcíaMarín et al. 2007
ZACATECAS		
espadín	*Agave angustifolia*	*Mezcal Reviews*
masparillo	*Agave maximiliana*	Gentry 1982, *Mezcal Reviews*
pulquero	*Agave salmiana*	*Mezcal Reviews*

Appendix Three

Human-Agave Symbiosis: Domesticated Agave Species Used in Mezcal or Other Distillates

Note: Because we recommend stronger reliance on a diversity of cultivated, domesticated agaves for making spirits until consumers are assured that wild agave species are not being overharvested or driven into extinction, this list can be a key tool in that transformation. Data compiled by Gary Paul Nabhan, drawing on research by many ethnobiologists.

Scientific name	Common Name of Plant	Name of Distilled Drink	States/Countries Where Grown for Agave Distillates
Agave americana subsp. *americana*	Arroqueño	Comiteco	Chiapas, Coahuila, Durango, Jalisco, Nuevo León, Oaxaca, San Luis Potosi, Tamaulipas
Agave americana var. *expansa*	Maguey	Doubtfully used in mezcal today, but perhaps in 19th century	Arizona, Sonora, Jalisco
Agave americana var. *franzosini*	Maguey	Ornamental only?	California

Scientific name	Common Name of Plant	Name of Distilled Drink	States/Countries Where Grown for Agave Distillates
Agave americana var. *marginata*	Maguey de chichimeco	Occasionally used in making mezcal	
Agave americana var. *oaxacensis*	Blanco, Cenizo, de Horno, de Rayo, Sierra Negra, Yavi cuam	Maguey Blanco, de Pulque, Sierra Negra	Oaxaca, San Luis Potosi
Agave angustifolia var. *angustifolia*	Garapato, Peruano	Peruano, Peruano	Jalisco
Agave angustifolia var. *deweyana*	Zapupe verde	Largely used for fiber	Tamaulipas, Veracruz
Agave angustifolia cv. "Espadin"	Espadín	Use in distillation of mezcal, possibly for fiber as well	Oaxaca, San Luis Potosi
Agave angustifolia var. *pacifica*	Mezcal bacanora	Used for mezcal alone (bacanora) or in blends	Sonora, Chihuahua
Agave applanata	Maguey de la casa, Maguey de Castilla, Maguey de ixtle, Maguey tepozco	Possible use for distillation, but primary use for baskets, cordage, ropes, and other craftswork	Puebla and Veracruz, but anciently diffused to Chihuahua, Durango, Querétaro, and Oaxaca
Agave cantala var. *cantala*	Maguey de cincoañero	Largely used for fiber	Indonesia, Philippines, Vietnam
Agave cantala var. *acuispina*	Cantala	Largely used for fiber	El Salvador, Honduras
Agave decipiens	Agave de Florida, False sisal	Not used	Florida, possibly Yucatán, Quintana Roo
Agave delameteri	Tonto Basin agave	Not used in distillate	Arizona
Agave desmettiana	Pineless Jade agave or Smooth agave		Oaxaca, Yucatán, Veracruz

Scientific name	Common Name of Plant	Name of Distilled Drink	States/Countries Where Grown for Agave Distillates
Agave fourcroydes	Henequén, Sisal blanco	Henequén mezcal	California, Oaxaca, Yucatán
Agave hookeri	Maguey Manso, Bravo	Occasionally used for mezcal	Michoacán
Agave inaequidens	Maguey bruto, Maguey alto	Used in Pulque and Mezcal distillates	Michoacán
Agave karwinksii	Cuishe, Cirial, Barril, Madre cuishe, Tabaziche, Tobaxiche	Distillate	Oaxaca, Puebla
Agave mapisaga var. *mapisaga*	Pulquero, incl. Listocillo and Tarimbaro	Used largely for aguamiel and pulque, may be in some mezcal blends	Coahuila, San Luis Potosi, Zacatecas, Guanajuato, Puebla
Agave murpheyi	Hohokam agave	Not used in distillate	Arizona, Sonora
Agave phillipsiana	Grand Canyon century plant	Not used in distillate	Arizona
Agave rhodacantha	Mexicano, Yocogihua	Mexicano	Sonora, Nayarit, Jalisco, Oaxaca
Agave salmiana ssp. *salmiana*	Maguey de Pulque, including Maguey Negro, Maguey Verde	Comiteco	Chiapas Guanajuato, Michoacán, San Luis Potosi
Agave salmiana var. *ferox*	Maguey Pulquero, Blanco, Bo'ta, Chalqueño, Cornudo, Gax mini. Grande, Mããxo, Mano largas, Mayeé, Mutha, Prito, Sha'mini Taxihuada, Tlacametl, Tsam'niuada, Maguey Verde, Xaminip	Used largely for aguamiel and pulque, may be in some mezcal blends	Oaxaca, Puebla

Scientific name	Common Name of Plant	Name of Distilled Drink	States/Countries Where Grown for Agave Distillates
Agave sanpedroensis	San Pedro agave	Not used in distillate	Arizona
Agave seemanniana	Biliaá, Do-be, Dobzan, DuabEl Maguey chato, Papalometl, Tobalá chato	Used for mezcal distillate	Oaxaca, Chiapas, Honduras, Nicaragua
Agave sisalana	Sisal	Used largely for fiber	Chiapas, possibly Yucatán, Africa
Agave tecta	unknown	Probably used as fiber	Guatemala
Agave tequilana	Tequilana azul	Used for tequila and some mezcals of 100% agave distillates	Guanajuato, Jalisco, Michoacán, Nayarit, Tamaulipas
Agave verdensis	Sacred Mountain agave	Not used in distillate	Arizona
Agave yavapaiensis	Page Springs agave	Not used in distillate	Arizona
Agave weberi	Maguey aguamilero, Maguey liso, Weber agave	Maguey aguamilero	

Appendix Four

Plants and Animals Used in Cured or Infused Agave Distillates, in Pechugas, or in Seasoned Salts

Spanish	English	Scientific name	Oaxaca	Sonora	Jalisco	Michoacán
FAUNA/ANIMALS						
Borrego/Cordero	Sheep/Lamb	*Ovis aires*	X			
Cachorra	Lizard	Phrynomisadeae		X		
Conejo	Cottontail rabbit	*Sylvilagus* spp.	X			
Cordoniz	Quail	*Callipepla* and *Cyrtonyx* spp.	X	X		
Culebra/Vibora de cascabel	Snake/Rattlesnake	*Crotalus* spp.	X			
Gallina ponedora	Chicken	*Gallus domesticus*	X			
Guajalote/Cocono/Pavo	Turkey	*Meleagris gallopavo*	X			
Iguana	Iguana	*Iguana iguana*				X
Jamón Ibérico	Hog	*Sus scrofa*	X			
Venado	Deer	*Odocoileus* spp.				X
FLORA/PLANTS						
Aguacate	Avocado	*Persea americana*				X
Almendra	Almond	*Prunus dulcis*				X
Anis	Yerbanis	*Tagetes micrantha*	X	X		
Anis de Estrella	Star anise	*Illicium verum*	X			

Spanish	English	Scientific name			
Arandanó azul	Highbush blueberry	*Vaccinium corymbosum*			X
Arroz	White rice	*Oryza sativa*	X		
Barba de elote/Seda de maíz	Maize/corn silk	*Zea mays*	X		
Bellota	Emory oak acorns	*Quercus emoryi*		X	
Cacahuate	Peanut/Groundnut	*Arachis hypogaea*	X		
Cacao	Chocolate	*Theobroma cacao*	X		
Café	Arabic coffee	*Coffea arabica*			
Canela	Ceylon cinnamon	*Cinnamomum verum*	X		
Cascara de citrica	Other citrus skin or zest	*Citrus* spp.	X		
Cascara de naranjo	Sweet orange skin or zest	*Citrus sinensis*	X		
Chabacano	Apricot	*Prunus armeniaca*	X		
Chile	Chili pepper	*Capsicum annuum*	X		
Clavo	Clove	*Syzygium aromaticum*			
Cuapantli, quapantli	Wild root of leguminous shrub used as medicine	unknown	X		
Ginjebra	Ginger	*Zingiber officinale*	X		
Guayaba	Guava	*Psidium guajava*	X		

Spanish	English	Scientific name	Oaxaca	Sonora	Jalisco	Michoacán
Higo	Fig	*Ficus carica*	X			
Maiz criollo	Maize/Corn	*Zea mays*	X			
Maiz criollo elote	Sweet corn	*Zea mays*	X			
Mango	Mango	*Mangifera indica*	X			
Manzana criolla de la sierra	Spanish cider apple, naturalized	*Malus* spp.	X			
Mezquite	Mesquite	*Prosopis* spp.	X			X
Piña	Pineapple	*Ananas comosus*	X			
Platano chino/rojo	Banana	*Musa x acuminata*	X			
Plantano macho el perrón	Banana	*Musa x acuminata*	X			
Polvo de mole	Ground chocolate	*Theobroma cacao*	X			
Rosa de Castilla	Rose of Castille or of Damascus	*Rosa* spp.	X			
Te de Limón	Lemongrass	*Cymbopogon citratus*	X			
Tejocote	Mexican hawthorn	*Crataegus pubescens*	X	X		
Tuna de nopal	Prickly pear fruit	*Opuntia* spp.	X			
Uvalama/Igualama	Chastity plant	*Vitex mollis*		X		
Uva pasa	Grape	*Vitis vinifera*	X			

			X
Yerbanis	Mexican tarragon	*Tagetes lucida*	X
Zarzamora	Elmleaf blackberry	*Vaccinium corymbosum*	X

INVERTEBRATOS/INVERTEBRATES INCLUDING LARVAE AND INSECTS

Escarabajo del maguey	Maguey scarab beetle	*Acanthoderes funeraria*	X
Gusanos blancos, meoculis	Tequila giant skipper	*Aegilae hespera*	X
Gusanos rojos	Red maguey worms	*Comadia redtenbacheri, Hypopta agavis*	X
Picudo de hocico en sisal	Sisal snout weevil	*Scyphophorus acupuntatu*	X

Appendix Five

Mezcalexicon: A Folk Vocabulary

A

Abocado: Neither a lawyer nor avocado fruit, *abocado* refers to the artificial "mellowing" (or doctoring) of *joven* mezcals to chemically infuse them with more color and flavor. Although it is a legally accepted practice for one of the classes of mezcal as defined by the current norm, it allows caramel coloring, artificial flavors, oak extracts, high-fructose sugar, and glycerin to "soften" the flavor and appearance of a mezcal to avoid the expense and time of aging it in a wooden barrel. In essence, a *joven abocado* is somewhat like a fake *reposado* or *añejo*.

Aguamiel: A Spanish term echoing old Indigenous terms that literally mean "honey water" or sweet sap upwelling from the meristem of a *maguey pulquero* to be drunk fresh or fermented into pulque. Aguamiel is still a popular fresh beverage in the altiplano of Hidalgo (where it is peppered with chiles), Querétaro, San Luis Potosi, Zacatecas, and in the Valley of Mexico. In some localities, the term is also used for the diluted juices of a roasted agave extracted from crushing in a *tahona* or with *mazos*. (See **mazos, tahona**) In a few places like Comitán de Domínguez, Chiapas, it is distilled into spirits called *comiteco*.

Alambique: A term derived from classic Arabic, *al-inbīq*, which refers to a particular copper-pot style of still originating in the Middle East or Central Asia. In many versions of this still, its lower evaporation chamber (set over a fire) is separated from its condensation chamber by a gooseneck. The invention of this ancient alembic distillation apparatus is apocryphally attributed to Sufi alchemist Jābir ibn Ḥayyān, who may have lived in Iraq up until 806 to 816. The alembic still has variants that became adapted for mezcal making across much of colonial Mexico, even though its presence may not be as ancient as either the pre-Colombian or introduced Asian-style stills surviving in remote mountainous areas along the Pacific coast of Mexico.

Al Granel: A (usually) low-quality *mezcal espadín* or *tequila* sold in bulk or aggregated from various producers. It is usually fermented with the help of sugars, packaged yeasts, or other additives, and distilled only once. The term is used to refer to cheap, wholesale quantities of mezcal produced in Guerrero, Jalisco, or Oaxaca that are the Mexican equivalents of "two buck chuck" wines.

Alquitara: The term was originally a synonym for the *alambique* still, but has since come to connote any slow distillation process of mezcal used in Mexico.

Alto: This term regionally refers to a folk variety of *Agave inaequidens* in Michoacán and adjacent; other folk varieties or synonyms for this wild species include *bruto* and *largo*.

Amole: From the Nahuatl *amulli*, a subgroup of agaves whose leaves, rhizomes, or hearts are rich enough in sapogenins to serve as soaps or as medicines. Although a few have been used for *mezcal espumosa*, a foamy or sudsy spirit, they are now a small percentage of all agaves used in distillates.

Añejo: This Spanish term for any "aged" or "time-seasoned" product legally refers to tequilas or mezcals aged in wooden barrels for a minimum of twelve months. The finest mezcals of this type are typically aged from eighteen months to forty-eight months. If the *añejo* is of 100 percent agave, it is usually aged for thirty-six to forty-eight months.

Apulco: At least fifteen trademarked artisanal mezcals are currently produced in Apulco, Jalisco, using both *espadín* and *tequilana azul* agaves.

Arroqueño: A Spanish folk taxonomic term for the Oaxacan landrace or *raza criolla* scientifically known as *Agave americana* var. *oaxacensis*. It is a gigantic wild maguey that takes twenty to twenty-five more years to mature in its natural habitat in the central valley of Oaxaca, but less when brought into cultivation. This slow growth accumulates earthy, herbaceous flavors and candied, smoky aromas when the mature heads are ultimately roasted. The mezcals distilled from this variety may feature flavors and fragrances of melon, chocolate, green beans, or vegetables, with a savory finish.

B

Bacanora: An agave spirit distilled from the northernmost populations of *Agave angustifolia* in the *zona serrana* of Sonora, and minimally in the adjacent states of Chihuahua and Sinaloa. *Bacanora*—from the Yaqui or Cahitan terms *baca*, "carrizo or other aquatic plants," and *nora*, "sloping ridge"—is named for a valley near Sahuaripa, Sonora. The terroir of the mezcal derived from this plant in the Valle de Bacanora was so distinctive that agave taxonomist Howard Gentry retained the name *Agave pacifica* exclusively for these Sonoran populations for decades before ultimately lumping it into the single-most widespread agave species in 1982. In 2000, *bacanora* was given a DO apart from the one already given to ancestral and artisanal mezcal. *Bacanora* remains the pride of bootleggers in Sonora.

Barbeo: The trimming, pruning, or cutting of the terminal spines of the agave leaves (*pencas*) aimed to make the head *(cabeza)* grow better. There are different techniques for the *barbeo*, including *barbeo de escopeta*, "shotgun pruning," to induce premature ripening and growth.

Barranco: The term used for pits prepared for the *tatemada*, "roasting," of mezcal in Sonora and Chihuahua. (See **horno, tatemada**)

Barril, Barrica, Barileto: Various names for barrels and kegs of different sizes and shapes; also a measure of volume of mezcal in some localities. The *añejo* and *reposado* mezcals and tequilas are typically aged in white oak barrels imported from southern Europe, although Mexican coopers called *toneleros* now make some barrels from other native oaks. Most barrels used for mezcal have capacities up to 200 liters (60 gallons), although less voluminous ones are used for smaller batches. *Barril* is also a folk variety of *Agave karwinskii*.

Batidor: The person who manually serves as the "beater" or "masher" of the roasted agave must (*mosto*) in the traditional process of making an agave distillate. Traditionally, a (usually) naked worker descended into the wooden vats or tubs where the must had been deposited. He used his hands and feet to beat the fibers from the mashed *piñas* to aid the fermentation process.

Bayusa: A Sonoran and Chihuahuan term for the edible flowers of agaves, especially those of the mezcal *bacanora, Agave angustifolia*.

Bermejo: A folk variety of landrace of *Agave sisalana* with a vermillion blush. It was historically mixed in with *tequilana azul* to make *vino mezcal de Tequila* but cannot be legally used today. It is however, sometimes included today in *asembleas* or *ensambles* of multiple agave species.

Biguata: A morphological term used in Sonora and Chihuahua for the heart or meristem of an agave, which, when roasted, forms a finer texture for eating and distilling than the more fibrous *penca* leaf tissue. (See **mezontle**)

Binguís, Bingarrote: Historically, these terms referred to two distinctive runs of aguamiel passed through an *alambique* still, as noted in Pineda's report on beverages of New Spain in the 1790s. Today, *binguís* is the ferment from the juice of *maguey pulquero* (*Agave salmiana*), from which *bingarrote* is distilled on a small scale in Guanajuato as an uncertified agave distillate.

Blanco: A white or *joven* mezcal, one of the classes of mezcal that has received no aging or infusion after distillation. Never aged in wooden barrels, it is merely allowed to rest in stainless steel tanks for up to sixty days before bottling. For tequila, it is also commercially termed *plata, plato,* or silver, and considered by some to be the sharpest or most strongly flavored of the types. **Blanco suave:** An unofficial term indicating that a *blanco* tequila has had extra aging or adulteration with *abogado* additives to smoothen its sharp taste.

Bola: Another term for the trimmed ball-like head or *cabeza* of an agave. (See **piña**)

Bronco: A folk variety of wild or feral *Agave salmiana* subspecies *crassispina*. It is smaller and has narrower leaves than other *salmiana* varieties but is used for both pulque and mezcal in semi-arid scrublands of the altiplano states of central Mexico.

Bruto: A folk name of wild *Agave inaequidens*, which grows in the subtropical scrublands and forests of Jalisco and parts of Michoacán. This agave is used primarily in *raicilla de la sierra* (where it is stored in aged wood blended with *Agave maximiliana*) and some rare 100 percent agave distillates.

C

Cabeza: The Spanish term for "head," used for the core of the agave rosette before it is trimmed into a *bola* or *piña*. In other parts of Mexico, it is referred to as the *coba, táhuta,* or *chicata*. The same noun is also used for the first run of agave distillate to come through the still, which is usually discarded (or used in cheaper *al granel* mezcals). This first run is also called the *puntas* or *chuqui*.

Cacalotentli: *Agave angustarium* of Guerrero, Michoacán, and Oaxaca, also sometimes called *cacaya*.

Cacaya: This is one of the many Indigenous terms for the edible flowers or blossoms of agaves. This term is used by both the Mixtecos and Popolacas of Oaxaca and Veracruz. Maguey Cacaya is also used as a folk name for *Agave kerchovei* of Oaxcaca and Puebla, or *Agave peacockii* of Puebla.

Caldera: This term is used by Oaxacan *maestros mezcaleros* for the lower one of the two stout clay pots used in *olla de barro* stills. A *caldera* typically holds 40 to 50 liters of the agave must, called *tepache* or *mosto*.

Canoa: An Arauca term meaning "canoe." It is a hollow tree trunk in which cooked agave is mashed with mallets. Larger *canoas* may also be used for fermenting cooked agave. *Canoas* is also a name for the coastal region of Jalisco, which has a distinctive tradition of mezcal distillation sometimes confused with the *raicilla de la costa* tradition.

Cántaro: A black, watertight ceramic jug that is cured and then used in the traditional process of aging mezcal in Oaxaca and Puebla. In Oaxaca, a trough-like leaf of a plant drains fresh distillate into a receptacle that is also called a *cántaro*.

Capada, Capona: This term refers to agaves whose flower stalks (*quiotes*) have been severed or castrated at the base (*capado*). This process—also called *desquiote*—allows the meristem to swell with sugars and flavors to season into a richer agave for distillation. *Maguey capón* is prized for its high sugar content and strong flavor.

Capitel: A less commonly used term for the *montera* of a still.

Carrizo: A type of river cane *(Phragmites australis* or *Arundo donax)*, used as conduit in clay-pot distillation. This bamboo-like tube can be employed in the *venencia* to measure and balance alcohol content.

Cenizo: Ashy-gray varieties or wild populations of several different varieties used in mezcal, including *Agave durangensis* and *Agave shrevei* in the Sierra Madre Occidental.

Chacaleño: A folk name for an *espadín*-like *Agave angustifolia* landrace from Tamazula in the westernmost sierra of Durango.

Chato: A folk name for *Agave seemanniana* and/or *Agave americana* in Jalisco. Also used as a historic synonym for the folk variety more frequently known as *sahuayo* that was once in *tequilana azul* fields and in distillates of *vino mezcal de Tequila*.

Chichihualco: A traditional mezcal from Chichihualco de los Bravos in the state of Guerrero, which is now infused with floral or fruity flavors of Jamaica (mango, *nanche*, and *tamarindo*) by the brothers Florencio "El Pato" and Baldomero Marino, and their children.

Chico aguiar: A folk term used by *mezcaleros* for one of several wild varieties of *Agave angustifolia* used for *raicilla de la costa* made on the Jalisco coast not far from Puerto Vallarta.

Chilocuiles: The infamous "red worms" drowning in the bottles of some Oaxacan mezcals—called *gusanos rojos, tecoles, chilocuiles,* or *chinicuiles*—are really larvae of two moths, *Hypopta agavis* or *Comadia redtenbacheri*. Their supple bodies can be eaten as a snack with your mezcal shooter, or ground with salt to adorn the rim of your margarita glass. They are also an ingredient in the seasoned *pulque curado* drink of Oaxaca and Puebla called *tecolio*.

Chingurito: A historic term for a class of spirits *(aguardientes)* produced in Mexico from either sugarcane or agave in the eighteenth and nineteenth centuries.

Chino: A folk name for *Agave cupreata* used to distill mezcal in Michoacán.

Churi: A northwestern Mexican term for a small but flavorful agave plant constrained by growing in a rocky environment, but one that may still produce flavorful spirits.

Cimarrón: A folk name for wild *Agave salmiana* subspecies *crassispina*, as well as a more general term for the wild-flavored agaves used in the distillation of mezcals and other spirits.

Cirial: Of the many folk varieties of landraces in the *Agave karwinksii* complex of Oaxaca and Puebla, this is one of the most varied sets of agaves used for mezcals in all of Mexico.

Coa de jima: A sharp, round-bladed tool much like a hoe used by *jimadores* to trim *(barbear)* or to harvest *(tumbar)* agaves.

Cogollo: The meristematic tissue forming the point of bud elongation into a *quiote* or agave inflorescence, also called *cobata* in northwest Mexico.

Cola: A term for the "tails" of the second or last run during agave distillation. While they are generally undesirable because they are low in ethanol and high in toxic methanol, they are still used to adjust the final alcohol content of mezcal. In essence, they are the distilled liquids condensed and collected toward the end of distillation, after the second and final "cut" has been made. Also called *mezcal floxo* in Pineda's report on agave beverages of New Spain in the 1790s.

Comiteco: A distilled beverage prepared from aguamiel (and/or sugarcane) in the town of Comitán de Domínguez, Chiapas; there are distilled pulques occasionally prepared on the altiplano of Central Mexico as well.

Común: The liquids that are condensed from the first distillation of agave juice, that when re-distilled, become mezcal. They are also called *shishe* and *ordinario*.

Corazón: The "heart" or middle portion of a run of agave distillates that is used to make mezcal or tequila.

Coyote: A folk variety of *Agave american* subspecies *americana* used for distilling mezcal in Oaxaca.

Cuelgue: An archaic Spanish term for the entire process of elaborating 100 percent agave spirits, derived from holding and fermenting *(colgando)* the agave juices in rawhide bags.

Cuernito: Literally "little horn," this is a container fashioned from a hollow cow or goat horn used to measure, taste, and drink agave distillates.

Cuescomate: A Nahuatl term from *cues comatl,* for *olla de barro* distillation.

Cuestecomata: The name for the hard-shelled fruit and tropical tree *Crescentia alata,* an early domesticated perennial of tropical America. The gourd-like fruits are painted or carved and used as decorative *jicarita* cups for sampling and sipping mezcals.

Cuishe, cuixe: A cylindrical, sometimes teardrop-shaped folk variety of wild *Agave karwinskii,* distilled in *olla de barro* stills for mezcal in Oaxaca. It is one of the few agaves that forms a woody trunk below the *piña,* which itself is elongated rather than round. Related to other folk varieties named *bicuishe* and *madrecuishe.*

Curandero: A folk name for wild *Agave marmorata,* also known as *pitzometl,* used for distilling mezcal in Oaxaca and Puebla.

E

Elixir de agave: A liqueur made with agave distillates—most often tequila and mezcal—infused with other liquors made from fruit or flowers, such as *damiana.*

Espadilla: A folk variety of *Agave angustifolia* var. *rubescens* used to distill mezcal in Jalisco and adjacent states.

Espadín: The cultivated variety of *Agave angustifolia* that is commonly used to produce *mezcal de Oaxaca,* but is also now grown in other states. It has become highly selected, cloned, and genetically narrowed just as the *tequilana azul* cultivar has become.

Estoquillo: A folk variety of *Agave univittata* var. *lophantha* (formerly *Agave lophantha*) used for distilling mezcal in Tamaulipas.

Excommun: An uncertified bootleg mezcal, also called *Excommunicación,* historically produced in Michoacán and still elaborated in one locality today. As described in Pineda's report on beverages of New Spain in the 1790s, it was prohibited by the Bishop of Valladolid, who sentenced culpable drinkers or distillers with excommunication and time in jail.

G

Giganta: A folk name for the rare and federally protected species *Agave valenciana*. It grows in the *raicilla de la sierra* area of Jalisco, where, despite its endangered status, it is authorized for use in mezcal distillation under the CRR. Fortunately, it is hardly used. Also called *relisero*.

H

Hijuelo: This Spanish term refers to rhizomes that extend from the base of mother plants, more often in some maguey varietals than others. They are natural genetic clones of the parent agave, and also called offshoots or "pups." *Jimadores* pull them up and replant vigorous ones in the first year or two of a new planting, but often cull out others that compete with the older mother plant for energy. In some areas, the size of *hijuelos* for replanting are classed by likening them to lemons, oranges, and grapefruits. They are also called seeds or *mecuates*.

Horno: The term for any kind of oven in Spanish, it is used by *mezcaleros* to refer to the above-ground brick ovens used by many producers to cook agave. It can also be used for underground ovens or baking pits, but *barranco* is more commonly used in Mexico's northwest and west central regions. In some regions, the entire *mezcalería* site for roasting, fermenting, and distilling agaves is referred to as *horno*, rather than as a *vinata*, *tren*, *taberna*, or *palenque*.

Huitzila: A mezcal made with the *tequilana azul* cultivar in the towns of Huitzila and Tezontla, Zacatecas.

I

Ingüixe: In some regions, this Indigenous term refers to the very end of the tails of distillation. Not used for adjusting the final alcohol content of mezcal.

Ixtle: This Hispanicized word from the Nahuatl term *ixtli* or *ictli* refers to agaves historically grown largely for the fiber use in ropes, handbags, baskets, and weavings. They are now used in the distillation of mezcal as well. *Ixtlero amarillo* is the folk name for a low-sugar variety of *Agave rhodacantha*, while *ixtlero verde* is a folk variety of *Agave angustifolia* in the same range of western Mexico. *Henequén (Agave fourcroydes)* is a fine fiber agave native to the Yucatán peninsula that *mezcaleros* grow for agave spirits in Yucatán, but it is distilled in Oaxaca. Sisal *(Agave sisalana)* is another agave originally domesticated for its fiber that is now used in *asemblea,* or blends of multiple species for mezcal or other 100 percent agave spirits.

J

Jabalí: A folk term for wild *Agave convallis* and the mezcals elaborated from them in Oaxaca and Puebla.

Jaibica: A small hatchet or *hachuela*, half the size of a normal axe, used specifically for trimming or mashing agaves in northwest Mexico.

Jaiboli: A rare wild agave of southern Sonora and adjacent Chihuahua. *Agave jaiboli* was historically used for distilling mezcal by Mayo, Guarijio, and mestizo inhabitants living on the flanks of the Sierra Madre Occidental. Legend has it that a campesino offered the name *jaiboli* for this agave to Dr. Howard Gentry, who codified it as a scientific epithet, before realizing that the well-traveled Sonoran meant that it should be good for "making highballs."

Jarcia: A term used in Tamaulipas to refer to several wild agaves—*Agave gentryi, A. montana,* and *A. montium-sancticaroli*—used for distilling mezcal in a distinctive tradition of northeastern Mexico.

Jicarita: Hispanicized from the Nahuatal *xicali,* a gourd-like fruit cask from the plant *Crescentia alata* used to measure, taste, and drink mezcal.

Jima: From the Nahuatl term *xima,* "to smooth into a desired shape." The term is used throughout much of Mexico for the pruning and harvesting of mature agaves.

Jimador: With roots as noted above, a skilled farmworker who harvests agave, most often used regarding tequila specialists.

Joven: One of the classes of mezcal as defined by the norm. White mezcal, receiving no treatment after distillation. Also called *blanco*, although *joven* is the traditional and preferred term.

K

Komil: An Indigenous term for "intoxicating drinks," supposedly derived from Nahuatl. It was a seldom-used name unsuccessfully proposed as a legal term for agave distillates that were not approved by the Consejo Regulador del Mezcal or other regulatory councils. Between 2015 and 2017, the CRM proposed NOM 199 that would have mandated that the Mexican government enforce the use of name *komil* for any agave spirits made outside the permitted zones for *mezcal*. The implementation of this change was never achieved, due to backlash from *mezcaleros*, who were furious that the CRM was determined to keep them from calling their plants and spirits *mezcals* as their families had done for centuries.

L

Lamparillo: Literally "little lamp," this term refers to a folk variety of *Agave asperrima* used for distilling mezcal in Durango. The folk term is also used to refer to a *trago grande* or big shot or swallow of any *aguardiente*.

Lechuguilla: Literally any wild plant with milky or cream-colored leaf sap (not just lettuce), this term is a folk name for several species of agave, not just *Agave lechuguilla*, a desert species first described in the Chihuahuan Desert in 1859. It is also the common name for *Agave maximiliana* used in *raicilla de la sierra*, and for *A. palmeri*, *A. shrevei*, and other related species distilled alone or mixed with *bacanora* in Sonora and Chihuahua. Also, *aguas de lechuguilla* is a mildly fermented probiotic favored by schoolchildren in Jalisco and Colima, sometimes produced from *A. inaequidens*.

Lineño: A folk variety of *Agave angustifolia*, also called *pata de mula*, once used along with *tequilana azul* in distilling *vino mezcal de Tequila* until banned by the CRT.

M

Madrecuishe: A folk variety of wild *Agave karwinskii* used in distilling mezcal in *olla de barro* stills in Oaxaca. (See **cuishe**)

Madurado en vidrio: Literally "matured in glass" (demijohns), this is one of the categories of aged mezcal as defined by the norm.

Maguey: From the Carib language of Hispaniola, this is now a preferred term for agave in Oaxaca and other states. The Spanish conquistadores brought the word with them from Hispaniola.

Maguey pulquero: A collective term used for the gigantic agaves such as *Agave atrovirens, A. mapisaga, A. salmiana,* and sometimes *A. americana* landraces. Many of these not only produce enormous quantities of aguamiel for fermentation into pulque, but can be distilled in 100 percent agave spirits such as *comiteco* as well.

Manso: A folk name for wild or feral *Agave salmiana* subspecies *crassispina.*

Margayate: Mezcals or agave spirits of crude quality, also called *broncos* or *soyate.*

Marsaparillo: A folk variety of *Agave lophantha* used in distilling mezcal in Tamaulipas.

Masparillo: A wild variety of *Agave maximiliana* occasionally harvested to distill mezcal in the Mezquital area of Durango.

Mayahuel: The Mesoamerican goddess of agaves, drunkenness, fecundity, and fertility. She dispensed aguamiel or pulque from her four hundred breasts to her children, the Centzon Tōtōchtin, or "drunken bunnies."

Mazos: Large wooden mallet used for hand-mashing cooked maguey in ancestral mezcals and other agave spirits.

Mechichicual: The Nahuatl term for the lateral spines on the sides of agave leaves, used to separate quality and ripeness of the *pencas.*

Mecuate: From the Nahuatl term *mecuatl,* the vegetative offshoots of a mother agave. (See **hijuelo**)

Meocuil: This Hispanicized Nahuatl term (from *meocuillin,* from *metl* and *ocuillin*) refers to the edible white larvae of the tequila giant skipper butterfly. It became a loan word in Mexican Spanish and is now widely used by *mezcaleros* everywhere. These larvae infest *tequilana azul* and other cultivated agaves.

Mescalón: A high-proof mezcal.

Metepantle: The agave-lined terraces of ancient Mesoamerican and Arid American *milpas*.

Metl: The Nahuatl word for agave, from which the word *mezcal* is derived: *metl ixcalli,* or "cooked agave."

Mexocotl: Spirits made with a wild, spiny bromeliad or *ciruela* botanically known as *Bumelia humilis.* Although the pineapple-like plant is not agave, its Nahuatl name means "fruit of maguey."

Mezcal, Mescal, Mexcalmetl: As noted above, Hispanicized Nahuatl terms for roasted agave, derived from *metl,* "agave," and *ixcalli,* "cooked," although there are other, more metaphorical etymologies as well.

Mezcal bronco: Mid-quality or mediocre mezcal, often for retail sale. Often purchased in bulk by companies for bottling.

Mezcal bruto: A folk variety of *Agave inaequidens* in Jalisco.

Mezcal casero: Small-batch, homemade (often bootlegged) mezcals or 100 percent agave spirits.

Mezcal colorado: Literally "reddish" or "rust-colored" mezcal juices. Oaxacan campesinos have a particular fondness for the rust-tainted distillates collected in old iron condensers.

Mezcal corriente: A term used for cheap, often adulterated agave spirits like Tonoyán *licor de agave,* which is used for *"aguas locas"* punches or cocktails the way Everclear grain alcohol or rectified spirit have been used. Historically, many decent mezcals were also considered *corrientes* unworthy of attention by sophisticated drinkers.

Mezcal curado: A mezcal or 100 percent agave distillate seasoned or infused with fruits, herbs like damiana, or nuts added after all distillation runs. *Pechuga* mezcals are infusions in one sense, though they are made through a slightly different process of letting vapors in a third run rise up through a bag of natural flavorants perched just below the condenser of the still.

Mezcal ticushi: From the Mixtec term, *Yavii ticunchi'i,* this folk name refers to *Agave nussaviorum* in Oaxaca. It is considered a sacred agave among the Mixtecs, who use the plant itself in rituals connected to the game *pelota*. There are also medicinal rituals associated with the very special mezcal distilled from this agave, distillates that are variously known as *el del patron* or *el del cura*, not just *mezcal ticushi*.

Mezcalería: A distillery, tasting room, or store for mezcals and other agave spirits.

Mezcalero, -a: The skilled artisan who oversees the baking and fermentation of *piñas* and the distillation of agave must in a *palenque, taberna,* or *vinata*. The most highly regarded of these liquid artists are called *maestros mezcaleros* or *palenqueros*.

Mezontle, Mesontle, Meylonte, Mosolote: From *metl*, "agave," and *zolotl*, "heart" or "meristem." The heart of the *piña*; it has a more granular texture and distinctive flavor profile compared to the *pencas*. On occasion, mezcals are distilled exclusively from *mezontles*.

Milpa: The agroecological landscape in which maize, amaranths, beans, squashes, chiles, and other vegetables are placed between rows of agave, prickly pear, and perennial tree crops. The *cosmovision* underlying *milpa* design favors crop diversity over uniformity, spatial heterogeneity over monoculture terrain, and multiple cultural, utilitarian, and spiritual uses of each plant over a single economic return.

Mistela por alambique: From Pineda's report on beverages of New Spain in the 1790s, an *ordinario* run drawn from an Arab-style *alambique* to be mixed with anise and sweet sap or syrup from pulque agaves (*necuhtli*).

Mixiote: This Nahuatl-derived term refers to the parchment-like cuticle or waxy membrane obtained from maguey leaves as well as to the barbecued meat dishes wrapped within it. It is a Hispanicized version of *metl*, for maguey or agave, and *xiotl*, the cellophane-like skin of its leaf.

Mixto: A kind of agave-based spirit—usually a cheap tequila—that is composed of fermented agave must combined with another source of sugar. Legally a *tequila mixto* needs to have at least 51 percent blue agave alcohol with 49 percent or less alcohol from sugarcane. *Mixtos* are used in most of the "industrial strength" margarita mixes used in bars and restaurants in the US.

Mochomos: Literally, this northern Uto-Aztecan term refers to the agriculturally industrious leaf-cutter ants (*Atta mexicana*) that ferment foliage into nutritious food for their brood in the dark recesses of underground burrows. Metaphorically, it is used in Sonora for bootleggers who distill *bacanora, lechuguilla,* and other spirits during the dark of night at stills hidden in shaded canyons.

Montera: A regional term used by *raicelleros* as a synonym for a wooden or metal *capitel* in stills used for making *raicilla* in Jalisco.

Moraleño: A folk variety of *Agave sisalana* once used in Jalisco for making *vino mezcal de Tequila,* but now banned in tequila production by the CRT.

Mosto muerto: "Dead must" derived from having fermented aguamiel into pulque or preparing it for distilling into *comiteco.* Essentially, it is a probiotic agave beer.

Mosto vivo: The "living must" of pulque, made from actively fermenting aguamiel with a wide variety of endemic yeasts and bacteria.

N

Necuhtli: A Nahuatl term used for honeybees and stingless bees, as well as the agave sweet sap molasses made from slowly boiling down aguamiel extracted from pulque agaves.

O

Olla: Another term used for *caldera* in Oaxacan stills used in the distillation of agave spirits.

Olla de barro: A style of still and Oaxacan tradition of distilling agave spirits also called *cuescomate,* a Nahuatl term from *cues comatl.*

Ordinario: Liquid condensed from the first distillation of maguey juice. When re-distilled, it becomes mezcal. (See **común, shishe**)

P

Palangana: The catchment bowl in an *olla de barro* still; the term more generally means a wash basin or catch basin. Also called *yuifana.*

Palenque: A mezcal distillery—in all of its dimensions—often part of the *mezcalero's* home or property. The work site was originally named after the underground roasting pit in which agaves are cooked. In other regions, the same kind of distillery may be called a *mezcalería, taberna, tanichi,* or *vinata.*

Palenquero: A mezcal producer, especially the person in charge of the operation.

Papalometl: A folk variety or landrace of *Agave cupreata,* although the same term may be used in other localities for the more diminutive *Agave potatorum* or related species with a spreading, curving butterfly-like shape. From the Nahuatl, *papálotl,* "butterfly," and *metl,* "agave."

Parcela, Potrero, Yunta: Terms used for the piece of land owned or rented for the production of agaves.

Pata de mula: A folk variety of *Agave angustifolia,* also called *lineño,* formerly grown with seven or eight other varieties for tequila before being banned by the CRT.

Pechuga: Meaning "(poultry) breast," this term refers to a traditional mezcal style originally from Oaxaca in which poultry, other meats, and/or local fruits and spices are placed in a cheesecloth bag below the condenser during the final distillation. The vapor passes through the bag and captures complex aromas in the condensation that features in special runs of mezcal. Often done with the same *mole* ingredients to be used in holiday feasts, this tradition has jumped to other states and stimulated innovation in the use of novel ingredients. Defined as the *"Destilado con pechuga"* as a special class under the norm for mezcals.

Penca: Possibly of Portuguese or Catalán origin, this term refers to spiky or spiny-side leaves, particular those of a maguey. Some *mezcaleros* enjoy making batches of agave spirits exclusively with the more fibrous but distinctively flavored *pencas* versus *mezontles.*

Perlas: Literally, "pearls," this term refers to the bubbles that form on the surface of agave spirits when its container is shaken. The formation of a certain density of pearls indicates that the spirits have achieved a strength between 45 and 55 percent alcohol by volume. Reading *perlas* is an essential art during the *venencia* or balancing of runs, so it has become a point of pride and mark of authenticity for many *mezcaleros.*

Petaquillas: An agave distillate infused or mixed with orange juice and cinnamon. These infused spirits are only sold locally in parts of Guerrero. They are occasionally used as sacraments in religious rituals.

Picado: In Oaxaca, this term is used for the cutting and harvest of ripened agave to prepare them for roasting, fermenting, and distilling.

Piña: Literally "pineapple," another succulent with morphological similarities to the agave, this has become the common name for trimmed heads of maguey harvested to produce nearly all agave spirits.

Pizometl, Pichomel: A folk variety of wild *Agave marmorata* used in the distillation of mezcal and sore throat syrup in several states, such as Puebla.

Potosino: A folk variety of wild *Agave salmiana* subspecies *crassispina* that is used for both pulque and mezcal in San Luis Potosi and adjacent states.

Potrero: Another term for a parcel or plantation of agaves, although terms like *rancho, campo, de agave, yunta,* and *huerta* may also be used regionally.

Pulque: A slightly foamy, milky probiotic beverage made from fermenting the sweet sap or "honey" known as aguamiel that wells up in the middle of an agave plant that has been "castrated" through the process called *desquiote.* This enormously popular beer-like beverage was first described in print by Hernan Cortés in 1524, and the volume of its consumption in Mexico far outdistanced distilled spirits from agave until after World War II. This nutritious drink remains popular as both unflavored and fruit-infused drinks in the southern Chihuahuan Desert, the altiplano, and south-central Mexico, including the capital city, where *pulquerías* were where men met to discuss politics in a state of mild inebriation for well over a century.

Pulquero: Both the name for pulque producers and varieties of agaves used for either pulque or mezcal, such as *Agave atrovirens.* In mezcals, this latter species carries lactose and fruity flavors, a chalky texture, and high minerality.

Puntas: The "points" or heads that come first in each run of distillation. Flavorful and high in alcohol, they are commonly used to adjust the final alcohol content of mezcal. They are sometimes consumed on their own.

Q

Quiote: This Hispanicized term for agave flower stalks comes from the Nahual *quiotl*, which can signify the stem, stalk, or emerging bud of an inflorescence, also called *cúburi* or *piri*. *Mezcaleros* stay alert to a change in agave leaf color and shape that signals the impending emergence of the *quiote*, for that is when storage carbohydrates are converted to simpler sugars so that the agave can rapidly send the stalk up to flower. If left uncut, the *quiote* may grow tall enough that its nectar-rich blossoms are within sight, smell, and reach of pollinating bats and hummingbirds. But by "castrating" the *quiote* in different manners through the *capona* or *desquiote* process, they can either capture the upwelling of *aguamiel* to ferment into pulque or halt the translocation of sugars so that the heart and leaf bases of a mature agave swell up to a greater volume. Some *quiotes* are peeled for the *mixiote* skins, then roasted in coals and eaten or fermented to make the highly valued *quiote* mezcals. (See **aguamiel, capona, pulque**)

Quitupan: A *vino mezcal de olla* produced in the pueblo of Quitupan in southeastern Jalisco as early as 1785.

R

Raicilla: A set of western Mexican mezcals lumped into two geographical units—one from the Jalisco and Nayarit coast (*raicilla de la costa*) made primarily from *Agave angustifolia* and *A. rhodacantha*, and one from inland mountain areas (*raicilla de la sierra*) made primarily from *A. maximiliana* and *A. inaequidens*. *Raicilla* is a generic term used for mezcals prepared from as many as six species of agave growing wild in Jalisco and adjacent Nayarit. However, the two primary regions of production recognized by the CRR—*costera* and *serrana*—produce distillates through different processing traditions and storage containers (plastic *tinacos* versus wood barrels) that taste very different from one another. The name *raicilla* now applied to both agave spirits supposedly refers to the "little roots" that bootleggers used to ferment as a ploy to avoid paying taxes and permit fees for harvesting the aboveground meristem and leaves of agave that fell under legal jurisdiction.

Relisero: A synonym for *gigante*, the wild but rare *Agave valenciana* permitted for use in *raicilla* distillation in the Mascota, Jalisco, area.

Reposado: From the Spanish term for being rested or "in repose." *Reposados*, also called *añejados* (aged) mezcals are stored in wood barrels for only two to nine months, or sometimes twelve months, less than the *añejo* mezcals. This seasoning process can legally be done with either 100 percent agave or mixed mezcals.

Resollano: A term used from the bottomless, sleeve-like *olla* perched above the *montera* in *olla de barro* distillation in Oaxaca.

Revoltijo: Literally "a mess," this term refers to a punch or infused liquor made from a mélange of prickly pear cactus fruit juice and the skin or bark of *timbre* (fernleaf acacia, *Acaciella angustissima*) in Puebla, Tlaxcala, and San Luis Potosi.

Ruqueño: A folk variety of *Agave americana* subspecies *oaxacensis* used in mezcal production in Oaxaca.

S

Saite, Sáhite: The still fibrous, chopped-up mash of roasted agaves, to be fermented and later separated into juice and bagasse.

San Martineo, San Martin: A folk variety of wild *Agave karwinksii* used in mezcal production in Oaxaca.

Serrano: A folk variety of *Agave americana* used in mezcal production in Oaxaca.

Shishe, Xixe: A regional name for the liquids condensed from the first distillation of maguey juice. When re-distilled, it becomes mezcal. (See **común, ordinario**)

Sierra negra: A folk variety of *Agave americana* subspecies *oaxacensis* used in mezcal production in Oaxaca.

Sikua: From a Tarascan term for mezcal used by the Indigenous Purépecha inhabitants in three municipalities edging Lago Patzcuaro, *sikua* is an agave distillate particular to Michoacán, made in this state before Michoacán was included in the mezcal DO. Other than a collective trademark used by *mezcaleros* in four Purépecha villages, the *sikua* nomenclature is most likely being left behind since Michoacán can now officially make mezcal.

Simple: The alcohol-rich liquid condensed from the first distillation run of fermented agave juice. The term is used in Jalisco and Nayarit for the first run of *raicilla*, but in other regions the terms *común* and *ordinario* are more frequently used.

Sinque: In Pineda's report on beverages of New Spain in the 1790s, an *aguardiente de pulque* made from both fermented aguamiel and sugarcane juices passed through a still.

Sotol: While not an agave, this plant is in the related genus of succulents called *Dasylirion*, or "desert spoons" in vernacular English. Trimmed heads of sotol are distilled in much the same way as agaves are distilled as mezcal. They are sometimes roasted and fermented in the same *horno* or *barranco*. The key difference between most agaves and *Dasylirion* is that the latter are multi-headed off the same trunk (as *Agave karwinskii* often grows), so that multiple harvests can occur from the same mother plants over many years.

T

Taberna: One of several names used by traditional distilleries for the elaboration of *raicillas* or other agave distillates.

Tacuachito: A nickname in northwestern Mexico for distillation runs of undifferentiated quality—neither good nor bad—that still need to be balled through the *venencia* process.

Tahitzingu: From Pineda's report on beverages of New Spain in the 1790s, a *mezcal corriente* fermented in animal skins, to which timbre and pulque are sometimes added before it is distilled.

Tahona: This term harkens back to the Arabic *aṭṭāḥūn[ah]*. It refers to the large stone wheel pulled by a horse or mule to crush cooked agaves used in mezcals, *raicillas, bacanora,* and a few tequilas. The 2-ton stone wheel rolls around in a circular stone basin called a *molino equipcio* or *molino chileno*. Although this is the preferred method for extraction of agave juices for fermenting *mezcal ancestral* or *mezcal artesanal*, some Oaxacan *mezcaleros* object to having to use this technique to certify their mezcal, since traditionally they used *mazo* mallets for the crush instead.

Tanichi: In Sonora, a modest-sized makeshift distillery for *bacanora* or other small-batch mezcals and sotols. It also means "store" today.

Tatemada: A term that simply means "the roasting," *tatemada* implies pit roasting or steaming underground instead of more industrial cooking or baking in brick ovens. It slowly transforms the inulins in the agave to simpler sugars for fermentation and distillation.

Tauta: A folk name for the diminutive wild *Agave parviflora*, whose heads and flower stalks are included in small batches of agave distillates from Sonora and adjacent Chihuahua.

Tecolio: A fermented beverage of Oaxaca and Puebla made with pulque and the maguey larvae—called *tecoles, chilocuiles,* or *chinicuiles*—then cured with honey, fruits, nopal, or *cempasuchil* marigold flowers. In Pineda's recipes on beverages of New Spain in the 1790s, the larvae are toasted and reduced to a fine powder, before being added to pulque.

Teolote: A folk name variously used for both *Agave maximiliana* and for wild *Agave marmorata*.

Teometl: A folk variety of *Agave atrovirens*.

Tepache: The fermented pulp and juices of agave *piñas* that forms the mash or must prepared for distillation. It is also the name of a native probiotic fermented drink like pulque, made in a clay pot with agave pulp and juice, infused with clove and cinnamon. Boiled barley and unrefined brown sugar may also be added later, as the mash is fermented for another two days. It may also be prepared with pulque that is mixed with honey and anise seeds, and then boiled into an anisette.

Tepantle: A form of terraced agriculture in Mesoamerica and Arid America that may include agave cultivation, but necessarily with plantings along the lip of the stone terraces or *trincheras*.

Tepemete: A folk name for wild *Agave angustifolia* distilled for mezcal or other agave spirits in Durango.

Tepextate, Tepeztate: Folk names for wild *Agave marmorata* that are distilled into mezcals with sweet vegetal notes, and hints of roasted maize kernels, bell peppers, and cilantro.

Tequilana azul: A homogeneous clonal cultivar of *Agave tequilana* that is the only agave allowed under the Mexican NOM for tequila.

Tezontle: A porous, scored vesicular volcanic stone used for making *tahona* grindstones for crushing roasted agaves and *molcajete* grinding bowls for crushing spices.

Tlahuelompa: A distillate made from the *tequilana azul* cultivar in the state of Hidalgo.

Tobalá: A folk name for rare, diminutive types of wild maguey that grow in the shade of nurse trees and boulders at high altitudes in Oaxaca state. The name may be applied to *Agave nussaviorum, A. potatorum,* or *A. seemanniana* when distilled into small batches of super-premium mezcals.

Tobasiche: A Zapotec term for a folk variety of a micro-endemic form of wild *Agave karwinksii* used to distill mezcal in *olla de barro* stills in Oaxaca.

Tonayán: Made in the industrial town of Tonaya in central-western Jalisco, this brand has called its product *de agave,* even though its primary component is cane sugar that is then caramelized to a golden hue in order to look like aged *tequila del oro.* It is reputedly the cheapest alcoholic beverage made in Mexico.

Torrecillas: Both mezcals and sotols have been produced at *vinatas* in Torrecillas, Durango, where historically *mezcal de Torrecillas* was commonly used to distinguish this agave distillation tradition from others.

Tren: A Sonoran term for the mobile makeshift stills used for clandestine productions of *bacanora, lechuguilla,* and other agave distillates by *mochomo* bootleggers.

Tripón: A relatively new folk variety of *Agave karwinskii* used for distilling mezcals in *olla de barro* stills in Santa Catarina de Minas.

Tristeza y muerte: A potent cocktail of bacterial, fungal, and viral pathogens that caused a pandemic among monocultural *tequilana azul* plantations beginning in the late 1990s.

Tuchi: A folk term for the first, very smooth, and delicious run of a mezcal or sotol drunk immediately as it comes out of the still in Nayarit and

Durango. It was described by explorer Carl Lumholtz, one of the first eth-
nographers who wrote about Huichol stills.

Tumbaderos: A regional name for field workers who harvest, trim, and clear
away the *piñas* of mature agaves to ready for transport and roasting.

Tumbayaquis: A Sonoran term for a *mezcal corriente* so potent and crude
that it could knock over a strong Yoeme (Yaqui) warrior.

Tuxca: An uncertified 100 percent agave distillate made in the pueblo of
Tuxcacueso in the Sierra de Amula on the state border between Jalisco
and Colima. It is often sold today with a sangria-like punch at roadside
stands in Tuxcacueso and other towns. It more broadly refers to one of
the most ancient and still dynamic traditions of agave distillates that is
now centered around the village of Zapotitlán de Vadillo. Ironically, these
extraordinary artisanal spirits do not fall within the legal framework of
the DO for mezcal, even though they originate in the probable cradle of
mezcal production. They are often produced as *asembleas* or *ensambles*
of three agave varieties, Cimarrón and Lineño of *Agave angustifolia*, and
Ixtero Amarillo of *Agave rhodacantha*.

V

Venencia: A term used for the hollow tube made from the native *carrizo*
cane plant *Phragmites australis* employed for siphoning mezcal and mea-
suring the density and duration of its *perlas*. The act of doing so is the
verb *venen ciar.*

Verde: Literally "green," a folk name for one of several varietals of *Agave
angustifolia* used for *raicilla de la costa* in Jalisco.

Vicho: The northern Uto-Aztecan name, possibly from the language of
the Opata, used historically for agaves or mezcals in the *zona serrana* of
Sonora and adjacent Chihuahua. The related term, *víchota*, means agave
juices derived only from the leaves, which some *mezcaleros* claim give a
richer flavor.

Vinata: Another regional term for distilleries for agave spirits. (See **mezcalería, palenque, taberna**)

Vinaza: The liquid waste derived from the tequila- and mezcal-making process, as distinguished from the solid, fibrous waste or *bagazo*. If properly managed, both can be used as soil amendments to improve fertility.

Vino mezcal: The historic term—still used in remote areas today—for agave spirits. Tequila was originally known as *vino mezcal de Tequila*.

Vino mezcal de Guadalajara: In Pineda's report on beverages of New Spain in the 1790s, a must of water, honey, and the roasted heads of agaves fermented in animal skins, then distilled.

X

Xima: The Nahuatl term from which the Spanish terms *jima* and *jimador* are derived.

Y

Yahui ndodo: A dessert made with clover and a rare agave known in Mixtec as *Yavii ticunchi'i*, the wild *Agave nussaviorum* of Oaxaca.

Yocogihua: A distilled agave spirit, much like *bacanora* of Sonora, that was produced near Alamos, Sonora, where it appears to have been made with two cultivated landraces of *Agave rhodacantha* introduced into the Río Mayo region in the nineteenth century. The Yocogihua distillery continued to produce mezcal from 1888 until 1985, despite the prohibition of distilling and selling mezcal in Sonora begun by Governor Plutarco Elías Calles.

Yunta: A regional name for land under agave plantations. (See **parcela, potrero**)

Z

Zapupe: A folk variety of *Agave fourcroydes* traditionally used for henequen fiber, but more recently for distilling *mezcal*.

Zihuaquio: A *mezcal artesanal de Zihuaquio* that continues to be made from *Agave cupreata* in Guerrero, where it is sometimes infused with coconut.

Zotol: An agave spirit distilled only from the lower meristem of the *Zotolero maguey* in the state of Puebla, and often infused with raisins or prunes.

Bibliography

Ángeles Carreño, Graciela. "Mezcal in Traditional Medicine." *Mezcal Arte Tradicional* 98 (2010).

Anonymous. "Demandan productores y expertos una autoridad que vele por cultura del mezcal." *La Jornada*, May 8, 2018. https://www.jornada.com.mx/2018/05/08/sociedad/031n1soc.

Arenstein, Noah. "The Problem with Pechuga." *Punch*, January 16, 2020. https://punchdrink.com/articles/problem-with-pechuga-mezcal/.

Bowen, Sarah. *Divided Spirits: Tequila, Mezcal, and the Politics of Production*. San Francisco: University of California Press, 2015.

Bowen, Sarah, and David Suro. "America's Role in Latest Fight for Small Mezcal." *Punch*, March 22, 2016. https://punchdrink.com/articles/is-america-part-of-the-problem-or-the-solution-for-mezcal-nom-199-and-nom-186/.

Brooks, David. "Grandes intereses comerciales buscan controlar la palabra agave: experto." *La Jornada*, April 16, 2016. https://www.jornada.com.mx/2016/04/16//opinion/014a1pols.

Bullock, Tom. *The Mezcal Experience*. London: Jaqui Small/Quarto Group, 2017.

Bye, Robert A., Don Burgess, and Albino Mares Trias. "Ethnobotany of the Western Tarahumara of Chihuahua, Mexico: I. Notes on the Genus *Agave*." *Botanical Museum Leaflets, Harvard University* 24 (1975): 85–112. https://www.jstor.org/stable/41762295.

Coleman-Derr, Devin, Damaris Desgarennes, Citlali Fonseca-Garcia, Stephen Gross, Scott Clingenpeel, Tanja Woyke, Gretchen North, Axel Visel, Laila Partida-Martinez, and Susannah G. Tringe. "Plant Compartment and Biogeography Affect Microbiome Composition in Cultivated and Native *Agave* Species." *New Phytologist* 209, no. 2 (2016): 798–811. https://doi.org/10.1111/nph.13697.

Colunga-GarcíaMarín, Patricia, Daniel Zizumbo-Villarreal, and J. Martínez Torres. "Tradiciones en el aprovechamiento de los agaves mexicanos: una aportación a la protección legal y conservación de su diversidad biológica y cultural." In *En lo ancestral hay futuro: Del tequila, los mezcales y otros agaves,* edited by Patricia Colunga-GarcíaMarín, Alfonso Larqué Saavedra, Luis E. Eguiarte, and Daniel Zizumbo-Villarreal, 229–48. Merida, Mexico: Centro de Investigación Científica de Yucatán, 2007. https://doi. org/10.13140/RG.2.1.5192.1441.

Colunga-GarcíaMarín, Patricia. "La desaparición de los mezcales artesanales tradicionales." *La Jornada,* January 21, 2012. https://www.jornada. com.mx/2012/01/21/opinion/021a2pol.

Colunga-GarcíaMarín, Patricia. "Se llaman mezcales y están hechos con agave: no engañen al consumidor. *La Jornada,* April 16, 2012. https:// www.jornada.com.mx/2016/04/16//opinion/014a1pol.

Colunga-GarcíaMarín, Patricia, and P. Ramos-Rivera. "Base de datos de nombres técnicos o de uso común en el aprovechamiento de los *Agaves* en México" (2006). https://www.snib.mx/iptconabio/resource?r=SNIB-CS007.

CONABIO (Comisión nacional para el conocimiento y uso de la biodiversidad). "Base de datos de nombres técnicos o de uso común en el aprovechamiento de los agaves en México" (2016). http://www.snib.mx/ iptconabio/resource?r=SNIB-CS007&v=1.0.

Cummins, Ronnie. "Agave Power: How a Revolutionary Agro-Forestry and Grazing System in Mexico Can Help Reverse Global Warming." *Regeneration International.* January 20, 2020. https://regenerationinternational. org/2020/01/20/agave-power-how-a-revolutionary-agroforestry-and-grazing-system-in-mexico-can-help-reverse-global-warming/.

Davis, Sarah C., June Simpson, Katia del Carmen Gil-Vega, Nicholas A. Niechayev, Evelien van Tongerlo, Natalia Hurtado Castano, Louisa V. Dever, and Alberto Búrquez. "Undervalued Potential of Crassulacean Acid Metabolism for Current and Future Agricultural Production." *Journal of Experimental Botany* 70, no. 22 (2019): 6521–37. https://doi. org/10.1093/jxb/erz223.

Eguiarte, Luis. E., O. A. Jiménez Barrón, E. Aguirre-Planter, Enrique Scheinvar, N. Gámez, J. Gasca-Pineda, G. Castellanos-Morales, and A. Moreno-Letelier. "Evolutionary Ecology of *Agave*: Distribution Patterns, Phylogeny, and Coevolution (An Homage to Howard S. Gentry)." *American Journal of Botany* 108, no. 2 (2021): 216–35.

Eguiarte, Luis E., Enrique Salazar, Jordan Golubev, and Maria C. Mandujano. "Ecologia y genética del maguey." In *Agua de las Matas Verdes: Tequila y Mezcal*, edited by José Luis Vera Cortés and Rudolfo Fernández, 183–213. Mexico, D.F.: Artes de México/INAH, 2015.

Eguiarte, Luis E., and Valeria Souza. "Historia natural del Agave y sus parientes: evolución y ecología." In *En lo ancestral hay futuro: Del tequila, los mezcales y otros Agaves,* edited by Patricia Colunga-GarcíaMarín, Alfonso Larqué Saavedra, Luis E. Eguiarte, and Daniel Zizumbo-Villarreal, 3–22. Merida, Mexico: Centro de Investigación Científica de Yucatán, 2007.

Ezcurra, Exequiel. "Las adaptaciones morfo-fisiológicos de los agaves a los ambientes áridos y su prospective agroindustrial." In *En lo ancestral hay futuro: Del tequila, los mezcales y otros Agaves,* edited by Patricia Colunga-GarcíaMarín, Alfonso Larqué Saavedra, Luis E. Eguiarte, and Daniel Zizumbo-Villarreal, 387–94. Merida, Mexico: Centro de Investigación Científica de Yucatán, 2007.

Farrell, Shanna, and David Suro. "Can a Group of Bartenders Save Mezcal?" *Punch,* April 30, 2015. https://punchdrink.com/articles/can-a-group-of-bartenders-save-artisanal-mezcal/.

Figueredo, Carmen J., Alejandro Casas, Patricia Colunga-GarcíaMarín, Jafet M. Nassar, and Antonio González-Rodríguez. "Morphological Variation, Management and Domestication of 'Maguey Alto' *(Agave inaequidens)* and 'Maguey Manso' (*A. hookeri*) in Michoacán, México." *Journal of Ethnobiology and Ethnomedicine* 10, no. 1 (2014): 1–12.

Figueredo-Urbina, Carmen J., Alejandro Casas, and Ignacio Torres-Garcia. "Morphological and Genetic Divergence between *Agave inaequidens, A. cupreata,* and the domesticated *A. hookeri*: Analysis of Their Evolutionary Relationships." *PLoS One* 12 (2017): e0187260.

Fish, Suzanne K., and Paul R. Fish. "Agave (*Agave* spp.): A Crop Lost and

Found in the US-Mexico Borderlands." In *New Lives for Ancient and Extinct Crops*, edited by Paul. E. Minnis, 102–33. Tucson: University of Arizona Press, 2014.

Fleming, Theodore H. "Nectar Corridors: Migration and the Annual Cycle of Lesser Long-Nosed Bats." In *Conserving Migratory Pollinators and Nectar Corridors in Western North America*, edited by Gary Paul Nabhan, 23–42. Tucson: University of Arizona Press, 2004.

Gagnier, Mary Jane. "Ceremonial and Culinary Uses of Mezcal." *Mezcal Arte Tradicional* 98 (2010).

García-Mendoza, Abisai J. "Distribution of *Agave* (Agavaceae) in México." *Cactus and Succulent Journal* 74 (2002): 177–88.

García, Domingo. *Mezcal: Un espirituoso artesanal de clase mundial*. Oaxaca de Juarez: 1450 Ediciones, 2019.

Gardea, Alfonso A., et al. *Bacanora y sotol: tan lejos y tan cerca; So Far, So Close*. Hermosillo: CIAD, 2011. https://www.ciad.mx/archivos/revista-dr/RES_ESP2/RES_Especial_2_07_Gardea.pdf.

Gentry, Howard Scott. "The Man-Agave Symbiosis." *Saguaroland Bulletin* 29 (1975): 73, 80–84.

Gentry, Howard Scott. *Agaves of Continental North America*. Tucson: University of Arizona Press, 1982.

Gil-Vega, K., M. G. Chavira, O. M. de la Vega, J. Simpson, and G. Vandemark. "Analysis of Genetic Diversity in *Agave tequilana* var. Azul Using RAPD Markers." *Euphytica* 119 (2001): 335–41.

Goguitchaichvili, Avto, Miguel Cervantes Solano, Jesús Carlos Lazcano Arce, Mari Carmen Serra Puche, Juan Morales, Ana María Soler, and Jaime Urrutia-Fucugauchi. "Archaeomagnetic Evidence of Pre-Hispanic Origin of Mezcal." *Journal of Archaeological Science* 21 (2018): 504–11.

Gonzalez-Elizondo, Martha, and Raquel Galván. "El maguey (*Agave* spp.) y los Tepehuanes de Durango." *Cactáceas y Suculentas Mexicanas* 37 (1992). Sociedad Mexicana de Cactología. Mexico, D.F.

Gschaedler-Mathis, Anne C., Benjamin Rodriguez Garay, Rogelio Prado Ramirez, and José Luis Flores Montaño. *Ciencia y tecnologia del tequila: avances y perspectivas*, seconda edición. Guadalajara: CIATEJ, 2015.

Gschaedler-Mathis, Anne C. *Panorama del aprovechamiento de los agaves en México*. Guadalajara: CIATEJ, 2017.

Gutiérrez González, Salvador. "Riqueza organoléptica de los agaves Mexicanos." In *Agua de las matas verdes: tequila y mezcal*, edited by José Luis Vera Cortés and Rudolfo Fernández, 215–42. Mexico, D.F.: Artes de México/ INAH, 2015.

Hernández, Maria. "La crisis del agave: la sobre explotación es una de las problemáticas más urgentes en el tema del mezcal." *El Universal*, June 2, 2017. https://www.eluniversal.com.mx/articulo/menu/2017/06/2/ la-crisis-del-agave.

Hodgson, Wendy C. *Food Plants of the Sonoran Desert*. Tucson: University of Arizona Press, 2001.

Howell, Donna J. "Pollinating Bats and Plant Communities." *National Geographic Research Report* 1 (1974): 311–28.

Jacques-Hernández, Cuauhtémoc, O. Herrera-Pérez, and J. A. Ramírez-De León. "El Maguey Mezcalero y la agroindustria del mezcal en Tamaulipas." In *En lo ancestral hay futuro: Del tequila, los mezcales y otros agaves*, edited by Patricia Colunga-GarcíaMarín, Alfonso Larqué Saavedra, and Daniel Zizumbo-Villarreal (2007): 287–317. Merida, Mexico: Centro de Investigación Científica de Yucatán, 2007.

Janzen, Emma. *Mezcal: The History, Craft & Cocktails of the World's Ultimate Artisanal Drink*. Minneapolis: Quarto Publishing/Voyageur Press, 2017.

Jiménez Vizcarra, Miguel Claudio. *La distilación temptrana (de Noviembre de 1576)*. Guadalajara, self-published limited edition, 2018.

Kitchen Sisters. 2014. "The Tequila Activist: A Conversation with David Suro of the Tequila Interchange Project." *The Kitchen Sisters*. June 26, 2014. http:// www.kitchensisters.org/2014/06/26/tequila-activist-a-conversation -with-david-suro-of-the-tequila-interchange-project/.

Leach, Jeff D., and Kristin D. Sobolik. "High Dietary Intake of Prebiotic Inulin-Type Fructans in the Prehistoric Chihuahuan Desert." *British Journal of Nutrition* 103, no. 11 (2010): 1558–61. https://doi.org/10.1017/ S000711451 0000966.

López, M. G., and J. Urias-Silvas. "Prebiotic Effect of Fructans from *Agave*,

Dasylirion and Nopal." *Acta Horticulturae* 744 (2007): 397–404. https://doi.org/10.17660/ActaHortic.2007.744.45.

Luna Zamora, Rogelio. *La construcción cultural y económica del tequila.* Zapopan: Universidad Guadalajara, 2015.

Maciel-Martínez, Jazmín, Eduardo Baltierra-Trejo, Paul Taboada-González, Quetzalli Aguilar-Virgen, and Liliana Marquez-Benavides. "Life Cycle Environmental Impacts and Energy Demand of Craft Mezcal in Mexico." *Sustainability* 12, no. 19 (2020): 8242. https://doi.org/10.3390/su12198242.

Madlulid, Domingo A. "The Life and Work of Antonio Pineda, Naturalist of the Malaspina Expedition." *Archives of Natural History* 11 (1982): 43–59.

Martinez-Salvador, Martin, Ricardo Mata-González, Carlos Morales-Nieto, and Ricardo Valdez-Cepeda. "*Agave salmiana* Plant Communities in Central Mexico as Affected by Commercial Use." *Environmental Management* 49 (2012): 55–63. https://doi.org/10.1007/s00267-011-9759-4.

McGovern, Patrick E., Fabian H. Toro, Gretchen R. Hall, Theodore Davidson, Katharine Prokop Prigge, George Preti, W. Christian Petersen, and Mike Szelewski. "Pre-Hispanic Distillation? A Biomolecular Archaeological Investigation." *Journal of Archaeology and Anthropology* 1, no. 2 (2019). https://doi.org/10.33552/OAJAA.2019.01.000509.

Medellín, Rodrigo A., M. Rivero, A. Ibarra, J. A. de la Torre, T. P. Gonzalez-Terrazas, L. Torres-Knoop, and M. Tschapka. "Follow Me: Foraging Distances of *Leptonycteris yerbabuenae* (Chiroptera: Phyllostomidae) in Sonora Determined by Fluorescent Powder." *Journal of Mammalogy* 99 (2018): 306–11.

Melgoza, Carlos, Frida Valdivia, and Rodrigo Cervantes (reporting the analyses of José Salazar Flores). "Hasta la sangre: los agroquímicos que habitan el curpo de campesinos en Jalisco." ZONADOCS. Guadalajara: Ciénega University Center (CUCIÉNEGA) of the University of Guadalajara, 2020. https://www.zonadocs.mx/2020/01/14/hasta-la-sangre-los-agroquimicos-que-habitan.

Meza, Martín P. Tena, Ricardo Ávila, and Rafael M. Navarro-Cerrillo. "Tequila, Heritage and Tourism: Is the Agave Landscape Sustainable?" *Food, Gastronomy, and Tourism Social and Cultural Perspectives* 49 (2018).

Nabhan, Gary Paul. "*Mescal bacanora:* Drinking Away the Centuries." In *Gathering the Desert,* conceived by Gary Paul Nabhan and Paul Mirocha, 37–50. Tucson: University of Arizona Press, 1985.

Nabhan, Gary Paul. "Finding the Hidden Garden." *Journal of the Southwest* 37 (1995): 401–15.

Nabhan, Gary Paul, and Theodore Fleming. "Conservation of New World Mutualisms." *Conservation Biology* 7 (1993): 457–59.

Nabhan, Gary Paul. "Producción tradicional del mezcal bacanora en Sonora: uno de los factores en el rompimiento de la relación entre murcielagos y agaves?" In *Resúmenes del Primer Simposio Internacional sobre Agavaceas.* Mexico, D.F.: Instituto de Biologia, UNAM, Mexico, 1994.

Nabhan, Gary Paul. "Stresses on Pollinators during Migration: Is Nectar Availability at Stopovers a Weak Link in Plant-Pollinator Conservation?" In *Conserving Migratory Pollinators and Nectar Corridors in Western North America,* edited by Gary Paul Nabhan, 3–22. Tucson: University of Arizona Press, 2004.

Nabhan, Gary Paul, Erin C. Riordan, Laura Monti, Amadeo M. Rea, Benjamin T. Wilder, Exequiel Ezcurra, Jonathan B. Mabry, et al. "An Aridamerican Model for Agriculture in a Hotter, Water Scarce World." *Plants, People, Planet* 2 (2020): 627–39.

Nabhan, Gary Paul, Patricia Colunga-GarcíaMarín, and Daniel Zizumbo-Villarreal. "Comparing Wild and Cultivated Food Plant Richness between the Arid American and the Mesoamerican Centers of Diversity, as Means to Advance Indigenous Food Sovereignty in the Face of Climate Change." *Frontiers in Sustainable Food Systems* 6 (2022): 840619.

Nebeker, Ryan. "The Sustainability Challenges that Threaten the Agave Industry." *Salon,* January 1, 2021. https://www.salon.com/2021/01/01/the-sustainability-challenges-that-threaten-the-agave-industry_partner/.

Niechayev, Nicholas, Alexander M. Jones, David M. Rosenthal, and Sarah C. Davis. "A Model of Environmental Limitations on Production of *Agave americana* L Grown as a Biofuel Crop in Semi-Arid Regions. *Journal of Experimental Biology* 70, no. 22 (2019): 6549–59.

Nobel, Park S. *Desert Wisdom/Agaves and Cacti: CO_2, Water, Climate Change.* New York: iUniverse, 2010.

Nobel, Park S., Edmundo Garcia-Moya, and Edmundo Quero. "High Annual Productivity of Certain Agaves and Cacti under Cultivation." *Plant, Cell, and Environment* 15 (1992): 329–35.

Nolasco-Cancino, Hipócrates, Jorge A. Santiago-Urbina, Carmen Wacher, and Francisco Ruíz-Terán. "Predominant Yeasts during Artisanal Mezcal Fermentation and Their Capacity to Ferment Maguey Juice." *Frontiers in Microbiology* 9 (2018). https://doi.org/10.3389/fmicb.2018.02900.

Núñez Noriega, L., and V. Salazar Solano. "La producción y comercialización de bacanora como estrategia de desarrollo regional en la sierra sonorense." *Estudios Sociales* 17 (2009): 205–19.

Patrón Esquivel, and César Agusto. *Mezcal: alimentos y bebidas de los pueblos Indígenas de México 2*. Mexico, D.F.: CDI, 2015.

Reyes, Vicente. *Camino Maguey*. Oaxaca: Arte-Sano Taller, 2021.

Rocha, Marta, Sara V. Good-Ávila, Francisco Molina-Freaner, Hector T. Arita, Amanda Castillo, and Abisai García-Mendoza. "Pollination Biology and Adaptive Radiation of Agavaceae, with Special Emphasis on the Genus *Agave*." *Aliso* 22 (2006): 329–44.

Romero, Fausto. "Of Wisdom and Eternity." *Mezcal Arte Tradicional* 98 (2010).

Ruiz-Terán, Francisco, Paulina N. Martínez-Zepeda, Sara Y. Geyer-de la Merced, Hipocrates Nolasco-Cancino, and Jorge A. Santiago-Urbina. "Mezcal: Indigenous *Saccharomyces cerevisiae* Strains and Their Potential as Starter Cultures." *Food Science Biotechnology*, 28, no. 2 (2018): 459–67. https://doi.org/10.1007/s10068–018–0490–2.

Serra Puche, Mari Carmen, and Jesus Carlos Lazón Arce. "Etnoarchaeologia del mezcal, so origen y su uso en México." In *Agua de las matas verdes: tequila y mezcal*, edited by José Luis Vera Cortés and Rudolfo Fernández, 23–42. Mexico, D.F.: Artes de México/INAH. 2015.

Stewart, J. Ryan. "Agave as a Model CAM Crop System for a Warming and Drying World." *Frontiers in Plant Sciences* 6 (2015): 684. https://doi.org/10.3389/fpls.2015.00684.

Suro Piñera, David. "Rumbo a la regionalización de las de." In *Agua de las matas verdes: tequila y mezcal*, edited by José Luis Vera Cortés and Rudolfo Fernández, 225–231. Mexico, D.F.: Artes de México/INAH, 2015.

Toal, Rion. "Olla de barro y mezcal: Pots of Clay and Mezcal." *Garland Magazine,* September 7, 2018. https://garlandmag.com/article/mezcal/.

Torres, Ignacio, Jorge Blancas, Alejandro León, and Alejandro Casas. "TEK, Local Perceptions of Risk, and Diversity of Management Practices of *Agave inaequidens* in Michoacán, México." *Journal of Ethnobiology & Ethnomedicine* 11 (2015): 61–74. https://doi.org/10.1186/s13002-015-0043-1.

Trejo, Laura, Verónica Limones, Guadalupe Peña, Enrique Scheinvar, Ofelia Vargas-Ponce, Daniel Zizumbo-Villarreal, and Patricia Colunga-GarcíaMarín. "Genetic Variation and Relationships among Agaves Related to the Production of Tequila and Mezcal in Jalisco." *Industrial Crops and Products* 125 (2018): 140–49.

Trejo-Salazar, Roberto-Emiliano, Luis E. Eguiarte, David Suro-Piñera, and Rodrigo A. Medellín. "Save Our Bats, Save Our Tequila: Industry and Science Join Forces to Help Bats and Agaves." *Natural Areas Journal* 36, no. 4 (2016): 523–30.

Valenzuela-Zapata, Ana. "A New Agenda for Blue Agave Landraces: Food, Energy, and Tequila." *GCB BioEnergy* 3 (2011):15–24. https://doi.org/10.1111/j.1757-1707.2011.01082.x.

Valenzuela-Zapata, Ana. "Raicillas, mezcales artesanales de Jalisco." In *Agua de las matas verdes: tequila y mezcal,* edited by José Luis Vera Cortés and Rudolfo Fernández, 131–46. Mexico, D.F.: Artes de México/INAH, 2015.

Valenzuela-Zapata, Ana, and Gary Paul Nabhan. *Tequila! A Natural and Cultural History.* Tucson: University of Arizona Press, 2003.

Valiente-Banuet, Alfonso., M. A. Aizen, J. M. Alcántara, J. Arroyo, A. Cocucci, M. Galetti, B. García, D. García, J. M. Gómez, P. Jordano, R. Medel, L. Navarro, J. R. Obeso, N. Oviedo, N. Ramírez, P. J. Rey, A. Traveset, M. Verdú, and R. Zamora. "Beyond Species Loss: The Extinction of Ecological Interactions in a Changing World." *Functional Ecology* 29 (2015): 299–307. https://doi.org/10.1111/1365-2435.12356.

Valiente-Banuet, Alfonso. "Producción de mezcal puede causar colapso del ecosistema agavero: ecólogo UNAM." *Aristegui Noticias* 20 April 2019 press release. https://aristeguinoticias.com/undefined/mexico

/produccion-de-mezcal-puede-causar-colapso-del-ecosistema-agavero
-ecologo-unam/.

Villanueva, Socorro, and Héctor Escalona-Buendía. "Tequila and Mezcal: Sensory Attributes and Sensory Evaluation." *Alcoholic Beverages* 12 (2012): 359–74. https://doi.org/10.1533/9780857095176.3.359.

Villegas, Paulina. "Graciela Ángeles, la mezcalirella Mexicana que pone el alto el nombre de las mujeres en esa industria." *Vogue México*. October 24, 2019. https://www.vogue.mx/estilo-de-vida/articulo/graciela
-angeles-mezcal-de-oaxaca.

Wilson, Iris H., and Antonio Pineda. "Pineda's Report on the Beverages of New Spain." *Arizona and the West* 5, no. 1 (1963): 79–90.

Xiong, Lili, Miranda Maki, Zhiyun Guo, Canquan Mao, and Wensheng Qin. "Agave Biomass Is Excellent for Production of Bioethanol and Xylitol Using Bacillus Strain 65S3 and Pseudomonas Strain CDS3." *Journal of Biobased Materials and Bioenergy* 8, no. 4 (2014): 422–28. https://doi.org/10.1166/jbmb.2014.1453.

Zandona, Eric. *The Tequila Dictionary*. London: Mitchell Beazley Imprint/ Octopus Publishing, 2019.

Zavala, Juan Carlos. "Uso del agua, el lado oscuro tras el 'boom' de la industria del mezcal en Oaxaca." *El Universal*. January 11, 2021. https://oaxaca.eluniversal.com.mx/sociedad/11–01–2021/uso-del-agua-el-lado-oscuro-tras-el-boom-de-la-industria-del-mezcal-en-oaxaca.

Zizumbo-Villarreal, Daniel, and Patricia Colunga-GarcíaMarín, and Alondra Alejandra Flores-Silva. "Pre-Columbian Food System in West Mesoamerica." In *Ethnobotany of Mexico: Interactions of People and Plants in Mexico*, edited by Rafael Lira, Alejandro Casas, and José Blancas, 67–82. New York: Springer, 2016.

Zizumbo-Villarreal, Daniel, and Patricia Colunga-GarcíaMarín. "La introducción de la destilación y el origen de los mezcales en el occidente de México. In *En lo ancestral hay futuro: eel tequila, los mezcales y otros Agaves*, edited by Patricia Colunga-GarcíaMarín, Alfonso Larqué Saavedra, Luis E. Eguiarte, and Daniel Zizumbo-Villarreal, 85–112. Merida, Mexico: Centro de Investigación Científica de Yucatán, 2007.

Index

Page numbers in *italics* refer to illustrations.